T0259679

Textile Science and Clothing Technology

Series editor

Subramanian Senthilkannan Muthu, Kowloon, Hong Kong

More information about this series at http://www.springer.com/series/13111

Subramanian Senthilkannan Muthu
Editor

Consumer Behaviour and Sustainable Fashion Consumption

 Springer

Editor
Subramanian Senthilkannan Muthu
Head of Sustainability
SgT Group and API
Kowloon, Hong Kong

ISSN 2197-9863   ISSN 2197-9871   (electronic)
Textile Science and Clothing Technology
ISBN 978-981-13-4597-5   ISBN 978-981-13-1265-6   (eBook)
https://doi.org/10.1007/978-981-13-1265-6

Printed on acid-free paper

This Springer imprint is published by the registered company Springer Nature Singapore Pte Ltd. part of Springer Nature
The registered company address is: 152 Beach Road, #21-01/04 Gateway East, Singapore 189721, Singapore

This book is dedicated to:
The lotus feet of my beloved
Lord Pazhaniandavar
My beloved late Father
My beloved Mother
My beloved Wife Karpagam and
Daughters—Anu and Karthika
My beloved Brother
Last but not least
To everyone working in the fashion
sector to make it SUSTAINABLE

Contents

Analysing the Consumer Behavior Regarding Sustainable Fashion Using Theory of Planned Behavior

Canan Saricam and Nazan Okur

Abstract Sustainable fashion, which is a term used for describing the clothing that is designed for longer lifetime uses, produced in ethical production systems using materials and processes that are not harmful to environment and the workers and which incorporates the fair trade principles with sweatshop-free labor conditions and makes use of eco-labelled or recycled materials, gained favor by the fashion industry specialists and the consumption market. Although the aspects of sustainable fashion regarding the selection of material, manufacturing process and supply chain were discussed by the researchers, the analyses examining the approach of the consumers towards sustainable fashion are quiet limited in number. Extended from Theory of Reasoned Action (TRA) with an additional dimension corresponding to a volitional control, Theory of Planned Behavior (TPB) was confirmed to be an effective model to predict the consumer intentions analyzing the three independent determinants of intention, which are "attitude towards the behavior", "subjective norm", and "perceived behavioral control" respectively. TPB was used to explain the consumers' approach towards either sustainability concept in general or sustainable products and processes in specific fields such as organic food consumption, green lodging, visiting green restaurants, purchasing luxury fashion produced in sweatshops, buying organic cotton apparel, using environmentally friendly transportation. Within the analysis, TPB model was used in original form or extended form with additional constructs self-construal, beliefs, ethical concerns or past experiences. In this study, the approach of the consumers' intention toward sustainable fashion purchasing was investigated using TPB where TPB model was applied in original form and extended form including the behavioral beliefs, normative beliefs and control beliefs affecting the determinants of the intention construct. After conducting an online survey among 339 participants in Turkey, confirmatory factor analysis and structural equation modelling were used to analyse the models. The findings showed

C. Saricam (✉) · N. Okur
Department of Textile Engineering, Faculty of Textile Technologies and Design, Istanbul
Technical University, Inonu Cad. No: 65 34437, Beyoglu, Istanbul, Turkey
e-mail: saricamc@itu.edu.tr

N. Okur
e-mail: okurn@itu.edu.tr

© Springer Nature Singapore Pte Ltd. 2019
S. S. Muthu (ed.), *Consumer Behaviour and Sustainable Fashion Consumption*,
Textile Science and Clothing Technology,
https://doi.org/10.1007/978-981-13-1265-6_1

1

that both models were validated and can be used to explain the approach of the consumers towards sustainable fashion products. According to both models, among the three constructs determining the intention was found to be attitude construct primarily, which shows the degree of favourable or unfavourable evaluation of the behaviour in question. The perceived behavioral control corresponding to the perceived ease or difficulty of performing the behavior was found to be less influential on the customer intention whereas the subjective norms defined as the perceived social pressure to perform or not perform the behavior was found to be influential more on the original TPB model. In the extended TPB model, whereas the attitudes were found to be influenced by behavioural beliefs, no relation was found between control beliefs and perceived behavioural control. Finally, the subjective norms and mormative beliefs acted as a single construct to be influential on the purchase intention.

Keywords Sustainability · Sustainable fashion · Consumer behaviour
Theory of planned behaviour

1   Introduction

Increasing population and diminishing natural resources enabled the sustainability concept to become much more important. Since it has the core purpose of meeting the needs of todays' generation without giving harm to that of the future generation, it was approached positively to sustainability concept by the countries, the governments, the companies and the consumption market in the end. Actually, sustainability is a term linking the economic development, the social development and the environmental protection, and therefore it is highly related with the activities established in the manufacturing industries and so the fashion industry. Because, the fashion industry actually designs, produces, markets and delivers products to meet the requirements of the end users and it works hand in hand with the textile and apparel producers at one end; with the distributors and the distribution channels on the other end. Within this respect, it is heavily influential on the economy considering the manufacturing and retailing industry; it influences the society because of ethical production and selling of fashion products; finally, it effects the environmental pollution by using many different types of chemicals and water especially in dyeing and printing stages of the production. Thus, the term sustainable fashion was adopted by the fashion industry to decrease the concerns about environmental protection and to satisfy the consumer who are sensitive to the environment and have special concerns regarding the ethical and social issues.

Regarding the sustainability concept, there are many studies analysing different aspect of the concepts which covered the environmental influences of the textile and apparel production, providing sustainability within the fashion supply and value chain, the ethical and social compliance aspects of the fashion production and selling. Nonetheless, the researchers gave different names to the concepts related with sustainability by making some limitations and specific definition. These studies made valuable contribution on the understanding of the consumers' approach toward sustainability and sustainable fashion products. Nonetheless, the number of studies investigating the approach of the consumers towards sustainability in fashion industry and analyzing the consumer behaviour are quiet limited in number. The existing studies analysed only one or two concepts of sustainable fashion such as the approach of the consumers toward the apparel products from organic cotton, the luxury fashion production in sweatshops.

Theory of planned behaviour (TPB) was applied in many studies some of whose specific focus was sustainability. Derived from theory of reasoned action (TRA) with the addition of the construct perceived behavior control (PBC), TPB stated that the intentions whose determinants are 'attitude towards behavior', 'subjective norms' and 'perceived behavioral control', are the immediate antecedent of the behavior. Although, it was not applied to analyze the consumer behaviour towards sustainable fashion products in the fashion industry, TPB was used to analyse the consumer behavior in specific fields such as green products, hotels, organic food and so on.

The purpose of this chapter is investigating the consumer behavior toward sustainability in fashion industry. The specific aims are analysing the different aspects of the sustainability in the fashion industry, the approaches of consumers towards this concept and modelling the consumer behaviour using theory of planned behaviour underlying the specific motivations lying behind.

Using sustainable fashion term throughout the chapter and modelling, the first section of the chapter covered the sustainability and the related concepts, the specific terms used in fashion industry to imply the sustainability issues. This section was followed with the summary of studies within the literature which describe the perspective of the consumers and companies toward sustainability. In the third section, the TPB and its main constructs were defined and it was followed with the parts dedicated to the description of the motivations and antecedents in the original and proposed extended models. The explanation of the constructs was related with the issues regarding sustainability. The third section ended up with the applications of TPB model in terms of investigating sustainability concept in the other fields and specifically the fashion industry. Finally, in the last section an empirical study to determine consumer behaviour towards the sustainable fashion products in Turkey was given for which TPB model was employed within the analysis.

2   Sustainability and Related Concepts

2.1   Sustainability Concept in Terms of Sustainable Development, Sustainable Consumption and Marketing

The term sustainability was firstly used in 1713 in the field of forestry with the meaning of never harvesting more than the forest yields in new growth. Then it drew attention of the economists for whom one of the major topic is the scarcity of resources. Nonetheless, it first gained reconginition as a policy concept where it was anounced in the Brundtland Report in 1987.

Although sustainability is a general concept and it can be created only if it is adopted and improved by all the people, the companies and the institutes serving the people, the countries, the governments and the globe as a whole, three requirements for the sustainability concept should be visited in order to understand its integrity from the perspectives of the consumer and consumption market.

First of all, the sustainability concept requires sustainable development which was defined by the World Commission on Environment and Development as "meeting the needs of the present without compromising the ability of future generations to meet their own needs" (WCED 1987, 8). Thus, sustainable development is related not only for their own generation of the consumers but also for the next generations. The sustainable development is possible with the production and the usage of sustainable products by balancing the ecological, economic and social dimensions. Within these respect, the sustainable products from green materials should be designed and produced in an environmentally friendly manner and they should be presented into the market considering the elimination of plastic bags at check outs, the reduction of CO_2 emissions, the compliance with the internal codes of good conduct (in relation to child labor), the improvement of employment practices (male/female wage parity, hiring of handicapped workers) (Lavorata 2014).

In parallel with the sustainable development, the sustainable consumption should be adopted by the consumers. The sustainable consumption can be achieved by "the use of goods and services that respond to basic needs and bring a better quality of life, while minimizing the use of natural resources, of toxic materials and emissions of waste and pollutants over the life- cycle, so as not to jeopardize the needs of future generations" (Norwegian Ministry of Environment 1994).

Finally, sustainable marketing should be in accord with the sustainable development and consumption. In fact the sustainable marketing practices, which can be described as the set of sustainable activities such as designing and implementing of marketing strategy that makes a net positive contribution to the society by producing a sustainability report and measuring the marketing success through a blend of financial, environmental, and social performance, should be implemented by the companies serving the consumption market (Ferdous 2010).

2.2   Related Concepts

Considering the definition and the requirements of the sustainability, it can be understood that sustainability is a general term with many aspects. Because of this, actually there are many studies and hence many related terms with the concept sustainability which are narrower than sustainability in some extent. Whereas some of these terms are reflecting the environmental aspects of the sustainability, the others are in in close connection with the social aspects. Actually, these different terms built the frame for the studies regarding sustainability in terms of the sustainable consumption, the marketing and specifically the consumer behavior.

Some researchers preferred to use green or ecological to reflect the environmental aspect of sustaianability. While the term 'green products' was defined as "products that will not pollute the earth or deplore natural resources, and [that] can be recycled or conserved" (Shamdasani et al. 1993), different terms related with the term green and the environmental aspect of sustainability became 'green marketing', 'environmental marketing', 'ecological marketing', 'sustainable marketing', 'greener marketing' and 'societal marketing concept' (Mostafa 2007). Even a term called 'green consumerism', which was alternatively known as 'eco-friendly', 'environmentally friendly', or 'sustainable' were proposed and adopted by some researchers (Kim et al. 2013).

The other researchers concentrated on the social aspect of the term sustainability. In fact, Shaw and Shiu (2002) underlined the point that, the increase in the environmental awareness resulted in the the green consumerism and even ethical consumerism in the end. The researchers distinguished the ethical consumerism from green consumerism claiming that the ethical consumers have additional concerns such as fair trade and armament manufacture.

Whereas the point determining the engagement in ethical or sustainable consumption were related with the social practice norms of the market by some researchers, the ethical consumption was stated to include any issues related with the fair trade principles such as the use of the organically grown materials and the organic processes, working practices in developing nations, and depletion of natural resources (Shaw and Riach 2011; Bray et al. 2011). The general definition of ethical consumer was that the ethical consumers are the ones, who consider the results of their consumption on the humans, animals or physical environment (McNeill and Moore 2015).

Actually, the social issues were stated to have power on the modern consumption culture as in the case of environmental issues since creating a sustainable society is also very important (Phau et al. 2015). Thus, the social aspects of sustainability began to be important for the consumers. Specifically, sweatshops became one of the issue being focused on since they are the work places which present poor working condition to the workers with long working hours and low wages and employ child labor (Shen et al. 2012). And the consumers nowadays are much more aware about the influence of the sweatshops on the health, safety and human rights (Phau et al. 2015).

The ethical concepts in the trade was also claimed to comprise the social and economical aspects of the sustainability concept. The fairly traded products, which were described as "the products which are purchased under equitable trading agreements, involving co-operative rather than competitive trading principles, ensuring a fair price and fair working conditions for the producers and suppliers" by Shaw and Shiu (2002), are counted within the related concepts of sustainability. Some researchers drew attention to the relationship between fair trade, environmental benefits and sustainability. The fair trade organization was proposed to be the one that encourages the producers to improve the environmental sustainability of the production by reducing the use of pesticides and fertilizers (Ozcaglar-Toulouse et al. 2006).

Considering the concepts listed above, it can be stated that sustainability concept was handled by many researchers employing one aspect or the other. Although there are some differences in terms of their meanings, they act as apart of a whole which is sustainability.

2.3 Sustainability Concept in Fashion Business and Sustainable Fashion

Sustainability concept was also favoured in the textile and apparel industry and the fashion consumption market. Nonetheless, the sustainability concept was discussed from different perspectives as above.

One aspect of sustainability, which was discussed more by the researchers, became the environmental sustainability including the issues such as 'organic fashion', 'organic clothing' and 'organic materials and fibers' used in the manufacturing of the apparel and fashion products. The 'organic fabrics' are usually produced from the herbal fibers such as organic cotton, hemp and bamboo for which no insecticides are used during growth phase and; the animal fibers such as wool, which are grown organically. The organic fabrics are not subjected to any dangerous or chemical processing, they are colored and printed with vegetable dyeing materials and inks and; no dangerous finishing material are applied such as stain proofing. The 'organic clothing' is promoted to be non-allergic, good in quality and stated to be taken as a long term fashion trend which can be styled according to everyday fashion. Within this regard, the 'organic fashion' is described as the fashion that deals with the presentation of the clothing and accessories which are produced with the minimum use of harmful materials and chemicals and by giving minimum harmful impact to the environment (Maloney et al. 2014).

Another concept regarding sustainability is the 'eco-fashion'. Different from the organic clothing and organic fashion, 'co fashion' is defined to be covering the clothing products, which are designed and manufactured to maximize benefits to human being and society as a whole. Within the production of eco-fashion and eco-products, the environmental impacts such as biodegradability and the usage of recycled materials are considered and environmentally responsible production processes such dyeing with the natural dyes are suggested to be used (Joergens 2006). Eco-fashion is stated to provide a sustainable consumption within the fashion supply chain, which enables the upstream of fashion supply chain processes from sourcing to production and distribution to be environmentally responsible within the aim of satisfying customer expectations. Actually, the fashion consumers compromise their needs of fashion clothing to be environmentally friendly within the eco-fashion consumption concept (Chan and Wong 2012).

Yet another concept regarding sustainability became the 'ethical fashion', which was described to point the fashion products being "high-quality and well-designed products that are environmentally sustainable, help disadvantaged groups and reflect good working conditions" (Domeisen 2006). Within this regard, the eco-conscious apparel consumption is stated to have the intention of purchasing the apparel products, which are made of recycled fibers, or from organic cotton; to select the clothing based on labor abuses; to care about the disposal of fashion products; to have negative attitudes toward buying fashion counterfeit goods, and; to follow ethical issues in fashion purchase behavior (Machiraju and Sadachar 2014).

Considering all the terms related with sustainability, it can be proposed to use the term sustainable fashion which reflect all the dimensions above. Actually sustainable fashion term was used by United Nations in 1972 and it was defined as a term used for describing the clothing that is designed for longer lifetime uses, produced in ethical production systems using materials and processes that are not harmful to environment and the workers and which incorporates the fair trade principles with sweatshop-free labor conditions and makes use of eco-labelled or recycled materials, gained favor by the fashion industry specialists and the consumption market. For this chapter, sustainable fashion is selected to be used for the sustainable products within the fashion market.

On the other hand, in the fashion industry and in terms of fashion products, there are ongoing discussions which state that the fashion products are not suitable to be sustainable where fashion means newness. In fact, the business concept fast fashion has many opponents at this point. Being demanded by young female consumers having little awareness of the social impact of their fashion consumption mostly, McNeill and Moore (2015) underlined the point that fast fashion trend which encourages purchasing new items continuously, seemed to conflict with the sustainability concept and the ethically sound production. But, many fast fashion brands using sustainability in their marketing claims are in opposition to this thesis. For instance, H&M group claims to provide affordable, good-quality and sustainable fashion for everyone regardless of their income. While they are presenting sustainable fashion collections to the consumers such as the conscious collection, which was produced from the organic materials or while they are inviting consumers to recycling activities by collecting garments in stores, they are stating that they focus on the United Nations Sustainable Development Goals and the SDG agenda, they aim to accelerate the transition from linear to a circular economy. Besides, they encourage their supplier to make more effective use of resources, to improve working condition by being in communication with the labor market (H&M website). Nevertheless, another business concept in fashion which is slow fashion seems to have more common practices with the sustainability concept and thus it receives positive approach from the defenders of sustainability. The slow fashion business concept is found to be much more related with the discussion of consumer ethics since since as in the case of slow food, it requires the consumer to question the established practices and worldviews and the economic models supporting the fashion production and consumption (Fletcher 2010). Being highly focused on the 'valuing and knowing the object' (Clark 2008) within the slow fashion practice, the process of raw material to finished product is suggested to be understood as part of the consumption experience (Manchiraju and Sadachar 2014).

Given all the explanations and the discussion about the conflict of fashion business and the sustainability concept, the fashion industry has high awareness at this point and try to act as responsible as possible for sustainable fashion products.

3 The Approach of Consumers Towards Sustainability and Sustainable Fashion

3.1 Consumers' Perspective

The consumers are known to have the power in sustainable development since they can request from the retailers to define the rules in regard to social practices such as trade unions, human rights at work, banning of child labor, good governance and transparency of management, while they can also demand products which meet environmental criteria (Lavorata 2014). Because of this reason, there are many studies within the literature which investigates the consumers' approach towards sustainability each putting forward the reasons lying behind the preference and avoidance of the consumers from the sustainable products.

The major question within analysing the consumer behaviour was that if the consumers will have willingness to pay a price premium for the sustainable fashion products or not. Actually some researchers concentrated on finding out what the price premium the consumers will be willing to pay for the sustainable fashion products. Casadesus-Masanell et al. (2009) established a study which stated that the customers of the Patagonia brand would be willing to pay more for the apparel items made from the organic materials. Similarly, Hustvedt and Bernard (2008) determined the price premium value, which the customers are willing to pay for the organic socks, as $1.86. According to study established by Miller (1992), a 10 percent price premium might not influence consumers' willingness to purchase eco-fashion but a 25–30% price premium is considered unacceptable (Chan and Wong 2012). Other than finding the premium amounts and percentages, different researchers revealed different findings. For instance, Roberts (1996) stated that the consumers do not want to pay a price premium especially when they think that eco-fashion is too expensive. The willingness to pay more was stated to be usually occurring when the value of the product exceeds the perceived value of it (Dean et al. 2012; Keh and Xie 2009). On the contrary, Mostafa (2007) stated that with the increasing green consciousness of the consumers which implies that their purchasing habits has an impact on the ecological problems, the consumers choose some of the products while they avoid the others even if the ecologically friendly products are more expensive than the others (Mostafa 2007). This dilemma was also put forward in the purchasing of the ethical fashion products. As in the case of organic products, the ethical fashion concept is also known to actually increase the demand of the consumers for high quality products, which are well designed, environmentally and socially sustainable. Nonetheless, although unethical manufacturing practices and the employment of the child labor has drawn attention of both the media and the consumers, the consumers are known to live usually the dilemma between purchasing low priced goods and ethical fashion products (Phau et al. 2015). Thus most of the studies investigating the willingness of the consumers to pay more for the environmentally friendly products found out that the consumers don't prefer to pay more for these type of products which meant, the customers don' want to make any sacrifices for eco-fashion products (Chan and Wong 2012).

Some researchers investigated if the store related attributes could be influential in the approaches of the consumers beside price. Chan and Wong (2012) examined in Hong Kong the relationships between store related attributes of eco-fashion and the consumption decisions of the eco fashion consumers and tried to find out if this relationship is dependent on the price premium of eco-fashion. They showed that store-related attributes of eco-fashion positively influenced consumers' eco-fashion consumption decision but this relationship can be weakened because of the price level. Store related attributes such as customer service, store design and environment, store's ethical practices, and shop convenience, are found to be critical for eco-fashion consumption decision by the other researchers also (Erdem et al. 1999) The ethical practices such as offering recycling service and recyclable products in stores are suggested to increase fashion consumers' perceived effectiveness of environmental protection (Roberts 1996).

Some researchers emphasized on the fact that the purchasing habits of the consumer could be influential on the purchase of sustainable products. For instance, Bray et al. (2011) stated that although price is one of the major parameters in decision making, the consumers who are making habitual type of purchase indicated that they could ignore the high prices. Besides, the other consumers claimed that they could pay more money for the products which were produced locally.

Culture was stated to be another parameter that determined the approach of the consumers towards sustainable fashion products. Some researchers stated that the globalized marketplace, the knowledge, attitudes and behavior show some differences across the cultures (Mostafa 2007; Johnson et al. 2004).

Nonetheless, the moral characteristics were stated to be more influential on the understanding of the consumers' approaches. Minton and Rose (1997) stated that the consumers having environmental disposition are more likely to have an intention in proenvironmental behaviors such as recycling, purchasing environmentally-safe goods, searching for environmental-related information, and buying recycled goods. But the authors added that moral obligation is much more influential on performing environmentally friendly behaviors than having a concern for the environment itself.

Since some studies put forward the negative approach towards sustainable consumption and sustainable products, these negativities were also investigated by the researchers. Within these studies, it was put forward that, fashion consumers did not prefer to purchase eco-fashion because of its characteristics such as poor quality, bad touch or being produced from uncomfortable materials. Even some of the consumers stated that they would buy eco-fashion only if there was no compromise on quality (Carrigan and Attalla 2001; Joergens 2006; Niinimäki 2010).

On the other hand, some studies showed that the negative attitude toward the sustainability and related concept was stated to be arisen because of being unaware of these products and; having lack of information about them. In some studies, the consumers stated that they are unaware of the availability of the eco fashion products and it is difficult to reach detailed information about them (Joergens 2006). Specifically focusing on the disposal of fashion items, Birtwistle and Moore (2007) claimed that most of the fashion consumers have lack of knowledge about the negative effects of the fashion industry on the environment.

Within this regard, it was pointed that since the environmentalism led to sustainable consumption as the awareness about the consumption related problems increases, the consumers are supposed to be more willing to purchase environmentally friendly products. In fact, it was found that understanding the ethical dimensions of the product enables the consumers to purchase the product and even increase the willingness to pay more for the product (Phau et al. 2015). For instance, McGoldrick and Freestone (2008) stated that out that consumers are willing to pay a premium of at least 6% for the "ethically assured" garments.

Some researchers drew attention to the fact that purchase behavior was not always the result of the purchase intention. Cowe and Williams (2000) named this lack of action as the "ethical purchasing gap" and the 30:3 syndrome based on the large-scale study conducted in the UK. According to this study, it was found that whereas 30% of consumers claimed they had ethical concerns only 3% made purchases. Similarly Bray et al. (2011) found out that there are differences between consumers' intentions to consume ethically, and their actual purchase behavior. The reason for this was asked to the consumers. Some consumers stated that the ethical purchasing gap occurred because their purchasing behavior are lacking of the ethical considerations but they felt themselves guilty after making the purchase. Other consumers claimed that they have a negative perceived impact on the image and the quality of the products and finally the rest indicated that their individual effort would not have any impact.

Within this regard, Bray et al. (2011) listed the reason for purchasing gap as the lack of information, the quality perception which can be either good or bad, the inertia in changing the purchasing behavior, seeing these activities as marketing ploy and expressing doubt for purchasing ethical produced by suppressing the guilt saying that one individual can not make a huge difference. In relation to these, Phau et al. (2015) added that the negative approaches from the consumers toward sustainability arises because of the preference to live in comfort zone instead of taking action. Moreover, the doubt was put forward as the key driving force that held the people back.

Considering the points above, it can be stated that although the consciousness of consumers are rising regarding the sustainability concept and sustainable products, there are still both positive and negative approaches towards these products which present some unknowns about the adoption process of sustainable fashion products by the consumers.

3.2 Companies' Perspective

Sustainability concept is adopted by the companies and the different businesses which are serving products to the consumption market. Actually, sustainable consumption help the corporate decision makers to apply stricter environmental regulation and to focus on preserving the environment (Paul et al. 2016). Although Aggeri et al. (2005) stated that the companies adopting the sustainable development express "a desire to be in compliance with social demands, in the knowledge that such compliance is merely symbolic and has little connection with the company's real business" (Lavorata 2014), Sheth et al. (2011) put forward that it is important for all the companies to orient themselves in a a customer-centric approach to sustainability if they want to prosper both in todays and futures environment.

Specifically, the textile and apparel manufacturers try to satisfy the consumers by producing eco-friendly fibers and fabrics (Brosdahl and Carpenter 2010). As an outstanding example, Patagonia uses 100% organic cotton and environmentally friendly fabrics which are chlorine-free wool, hemp, recycled nylon and polyester and tencel lyocell made from sustainable eucalyptus trees. The producers take some precausitions such as the usage of little or no chemical within the products, the usage of non-toxic ingredients during the coloring and processing stages, not establishing chemical treatments during finishing (Maloney et al. 2014). Although fast fashion was stated to be nonsustainable, the sustainability practices implemented in the fashion industry showed that most of the fast fashion producers are using sustainable fibers (hemp, organic cotton, bamboo, post-consumer recycled fabrics), encouraging the reuse and second hand clothing channels, and supporting ethical labour practices (McNeill and Moore 2015).

Other than making some changes within their production processes and products, the companies adopt green marketing strategies by integrating the environment into their marketing mix and marketing activities including the package design, pricing and even by implementing green advertising and green marketing strategies (Mostafa 2007). For instance, the companies have begun to position their products on the basis of environmental appeal such as recycled paper, plastic goods and so on (Banerjee et al. 1995).

Arguably, Schaefer (2005) states that the sustainable development is a very challenging for the marketer where there are the global issues of overpopulation, increasing energy demand, loss of biodiversity, and the predicted wide-ranging negative impacts of climate change. As a matter of fact, the marketing and the concept of the sustainability has contradictory purposes. This is because whereas marketing encourages consumption, sustainability is meeting todays' generation needs without compromising the ability of the future generation meet their needs (Ferdous 2010). Nonetheless, the definition of sustainable marketing "a form and purpose that makes a net positive contribution to society in terms of environmental, social, and economic development" (Ferdous 2010) proves that sustainability in marketing is possible.

Although there is no clarity accepted by all the parties regarding the sustainable marketing, some tangible advises are also given to the retailers by the researchers studied on this topic. For instance, considering that the fashion consumers have physical, emotional and psychological needs, it was suggested to the fashion companies to improve their store related attributes to satisfy those needs as a marketing strategy to increase eco-fashion consumption and to facilitate the development of sustainable fashion supply chain (Lai et al. 2012). Moreover, it was stated that the garment tags and product packaging are considered by the consumers because of the increasing concern about the environmentally friendliness. And the marketing claims of the companies (marketing claims using the terms, such as eco- or environmentally friendly, ozone-friendly, recycled, recyclable, or reusable, without providing specific information about their meanings) become important (Phau and Ong 2007). The companies begin to apply environmental strategies using special brand names and logos, advertisements, and websites (Kim et al. 2012).

Besides the tangible ones, there are also some intangible advices for the retailers in order to encourage consumers to make sustainable consumption. It is suggested to relate the sustainable development with the positive values. Regarding this, since sustainable development is an elastic concept for which different meanings are given by different people, some companies are advised to keep the sustainable development equal to durability whereas the others are advised to include corporate social responsibility within this concept (Lavorata 2014).

Moreover, it was stated that the maximization of sales and consumption of green products can only be done by creating a shared sense of responsibility for the environment (Chen and Peng 2012). The shared sense of responsibility enables the consumers to adopt greener lifestyles in the long run (Paul et al. 2016). The shared sense of responsibility can be created by positive information but it was stated to be arisen by also announcing negative news. Because, some researchers claimed that the negative news have higher impact than positive information (Herr et al. 1991).

To conclude, the companies are highly aware of the significance of sustainable fashion and they are trying to produce some solutions and make some improvements. And these policies and the solutions can be applied on either value based or behavior based approaches whereas the consumer values can be tried to be emphasized in the value based solutions, some social and institutional changes can be made in behavior based solutions (Thøgersen and Ölander 2002).

4 Theory of Planned Behavior

The Expectancy value models are used mostly in order to predict and understand the consumer behaviour such as norm activation model (Schwartz 1977), motivation protection theory (Rogers 1983), the health belief model (Rosenstock 1974), social cognitive model (Bandura 1986), the theory of reasoned action (TRA) of Ajzen and Fishbein (1980), and Theory of Planned Behaviour (TPB) (Ajzen 1991).

As one type of expectancy value model, TRA is a social-psychological model where the person's actual behaviour in performing a certain action is directly guided by his or her behavioural intention which is determined by the subjective norm and attitude towards the behaviour (Fishbein and Ajzen 1975). Within TRA, Attitude refers to "the degree of a person's favourable or unfavourable evaluation or appraisal of the behaviour in question" (Fishbein and Ajzen 1975). Subjective norm, which are related with the normative social beliefs (Hansen 2008), refers to "the perceived social pressure to perform or not to perform the behaviour" (Ajzen 1991).

The main critic of TRA is that the behavior is perceived to be entirely voluntarily (Liao et al. 2007) and it performs badly for the people having incomplete volitional control. Thus, the construct perceived behavioural control (PBC) was included within the model of TRA to form TPB which can be regarded as "people's perception of ease or difficulty in performing the behaviour of interest" (Ajzen 1991, 2002); or "should be read as perceived control over the performance of a behaviour" (Ajzen 1991, 2002). Thus TPB differs from TRA because of the addition of PBC which is conceptualized as the belief of the consumer about the difficulty of the behavior which requires some skills, opportunities, risks and resources but don't actually occur because the consumers decide to act (Hansen 2008).

Thus, TPB model is composed of four constructs which are intention, attitude, subjective norms and perceived behavioral control. According to TPB, the behavioral intention, which was defined as "an individual's likelihood of engaging in the behavior of interest" is the direct antecedent of the actual behavior (Ajzen and Fishbein 1980; Pookulangara et al. 2011). There are some different opinions from the researchers regarding the intention however. The attitudes are known to be important predictor of behaviors, behavioural intentions and they act as an explanatory factor for individual behaviors in social psychology literature (Kotchen and Reiling 2000). The judgement about whether the behavior under consideration is good or bad is included within the construct attitude since it is the psychological emotion associated with consumers' evaluations. In TPB, it was stated that subjective norms involve the feelings of an individual regarding the social pressure felt about a given behavior, and the past experiences and anticipated obstacles are reflected by the construct PBC (Paul et al. 2016).

The Theory of Planned Behaviour (TPB) is an important social cognitive model for analyzing the volitional behaviour of the consumers. It was used in health psychology and health behavior (Kidwell and Jewell 2003; Walker et al. 2001; Watson et al. 2014; Zemore and Ajzen 2014) in analyzing the food choice and consumption (Vermeir and Verbeke 2008; Bissonnette and Contento 2001) in consumer satisfaction (Liao et al. 2007) and in apparel consumption (Kim and Park 2005). Moreover, TPB was used favorably in terms of analyzing the consumers proenvironmental behaviors (Bonnes and Bonaiuto 2002; Abrahamse and Steg 2009; Whitmarsh and O'Neill 2010; Chao 2012; Oreg and Katz-Gerro 2006; Harland et al. 1999), recycling behaviors (Paul et al. 2016; Davis et al. 2006; Chan 1998) and sustainable consumption behavior in general (Sparks and Shepherd 1992; Fielding et al. 2008; Maloney et al. 2014; Kalafatis et al. 1999; Mancha and Yoder 2015; Mostafa 2007), sustainable product consumption in different fields such as selection of the organic food (Arvola et al. 2008; Zhou et al. 2013; Yazdanpanah and Forouzani 2015; Tarkiainen and Sundqvist 2005; Hansen 2008), preference of the green restaurants (Kim et al. 2013), transportation (Heath and Gifford 2002), hotels

(Chen and Tung 2014; Han et al. 2009, 2011; Kim and Han 2010) and purchasing the organic apparel eco fashion products (Hustvedt and Dickson 2009; Jang et al. 2012).

4.1 The Beliefs and Thoughts Lying Behind the Theory of Planned Behavior

The TPB model was discussed and investigated by many researchers in order to understand its efficacy to analyse the consumer behaviour. Different cognitive and affective perspectives were highlighted for both TPB and its usage in terms of under-standing sustainability concept and sustainable consumption. Within this section, these will be summarized to consider the real motivations lying beneath the con-sumer behaviour.

TPB model is stated to be the cognitive accounting of beliefs and evaluations. Tikir and Lehmann (2011) underlined the fact that beliefs are related with the information about a specific object or a consequence of behaviour and they are influenced from cultural, personal, situational factors like the personality of the individual and the values he/she has. Thus, the personal characteristics and personal perception can be influential on the beliefs toward a specific behaviour regarding sustainability concept.

Related with the values, Wiseman and Bogner (2003) identified environmental values which have the dimensions; preservation (bio-preservation) corresponding to the conservation and protection of the environment and; utilization (anthropocentric utilization) corresponding to the utilization of natural resources. Within this respect, it is stated that biocentrics tend to feel more responsibility towards environment and they act more pro-environmental. In fact, the beliefs regarding the relationship between the self and the nature are the core elements and ecological identity, and eco-psychology (Mostafa 2007). According to Kals et al. (1999), the beliefs reflect themselves on the emotional bonds towards nature which deploy a motivation to preserve it.

Nonetheless, the personal characteristics and the perception of self is very related with the belief about sustainability. People may behave differently in accordance with describing themselves to have internal or external control. The consumers with external control indicate that they are making ethical purchasing because of external control; whereas the consumers with internal control indicate that they make ethical purchases if there is no conflict of the social and situational pressures (Bray et al. 2011).

The consumers are stated to behave differently depending on having egoistic or altruistic characteristics. Stating that each individual can position himself/herself on a value continuum from egoistic to altruistic and from conservative to open to change, the authors Corraliza and Berenguer (2000) revealed that the individuals, who are altruistic are the ones who adopt pro-environmental behaviour. Altruism, as "self-sacrificial acts intended to benefit others regardless of material or social outcomes for the actor" (Schwartz and Howard 1984) was depicted to be one of the major values which is related to socially responsible apparel consumer behaviour, who makes ethical consumption. Nonetheless, it was found that altruism is not causing an intention to buy apparel products from socially responsible businesses (Hustvedt and Dickson 2009).

Moral norms are also shown to be influential on the intention of the consumers since they reflect the influence of altruism behaviour. Schwartz's theory claims that pro-environmental actions occur because of the personal moral norms and the individuals believing in the fact that environmental conditions pose threats to other people, other species, or the biosphere, perform actions to avert those consequences (Stern et al. 1999). Moral decisions are described with the situations where the people are aware that the well being of the other people are dependent on their action so they feel responsible for both the action and its consequences (Arvola et al. 2008). The term integrity is stated to be associated with the moral values actually. Integrity is defined as having strong moral values (Phau et al. 2015) and the purchase intention is known to be influenced by integrity (Laeequddin and Sardana 2010). The people having high integrity prefer to make ethical purchases more than the people having low integrity (Floyd et al. 2013). Moreover, trust for that, buying the product will provide benefits increases the integrity, leads the consumers to purchase intention (Phau et al. 2015).

Other than the moral norms, social norms present a pressure for the consumer which increases according to the given importance to the others' opinions (Horng et al. 2013). So they can be regarded as the beliefs and values from the other people or groups. Nonetheless, the social norms or the social influence can be also attached with susceptibility to interpersonal influence as Phau et al. (2015) stated. The first type of susceptibility is the informational susceptibility which is because the people have to be more skeptical if they obtain the information from others. But the opinions of especially the knowledgable people can help the consumers to make decisions by acting as a reference points. The second type of susceptibility is the normative susceptibility, which is the tendency to comply with the expectations of others with a concern to impress them. This comes usually with a desire to enhance self image by association with a reference group. Since what the other people believe to be true is influential on the people's decisions, a person might observe the other peoples behaviors believing that these observations reflect the reality about sustainability concept (Phau et al. 2015; Hermans et al. 2007; Tang and Farn 2005).

Some of the researcher thought that personal norms should also be considered in order to analyse the behaviour. Actually, the subjective norms which are already included in TPB model, were stated to be composed of the normative beliefs. Subjective norms are related with the influence of the others on the individual with regard to performing a certain behavior (Minton and Rose 1997) and they are created according to how one individual thinks that the other view this individual (Maloney et al. 2014). On the other hand, personal norms are related with the individual's own perception regarding performing a certain behavior and thus they actually refer to beliefs of the person about the actions. Personal norms are found to be influential on one's product choice, information search and recycling habits (Minton and Rose 1997).

Some researchers preferred to use ethical obligation for indicating the personal norms. The claim regarding the ethical obligation is that, they actually represent the internalized ethical rules of an individual and they correspond to the personal beliefs about the right and wrong (Ozcaglar-Toulouse et al. 2006).

Another suggestion is that self identity can be also influential on the consumer behaviour. Self identity can be described as how one perceives oneself (Yazdanpanah and Forouzani 2015) and "reflect the extent to which a person sees him/herself as fulfilling the criteria for a particular societal role" (Pelling and White 2009). Hustvedt and Dickson (2009) claim that the self identification of the consumer as a green consumer or organic consumer could influence on the behavior of the consumers as in the case of their attitudes, beliefs and moral obligation to buy organic products. Self identity is stated to include all the roles of the individual affecting the behavior which may precede or contribute to expectations and norm since the choices are defined by the degree a behavior is consistent with one's sense of self (Stryker 1968). Mancha and Yoder (2015) add that one aspect of identity is self construal which means how people feel and think about themselves.

The relation between the self identity and showing collectivist or individual-istic belief regarding cultural heritage are considered by some researchers too. It is stated that self construal is influential on the socially responsible behavior espe-cially in the collectivist cultures where interdependence is much more important than independence (Mancha and Yoder 2015). But independence predicted the egoistic environmental concern (Arnocky et al. 2007). Markus and Kitayama (1991) adds that collectivists are much more influenced by normative beliefs while individualists prioritize personal goals, and more influenced by individual attitudes and perceived control. Although self is shaped with both the culture and experience, Oreg and Katz-Gerro (2006) implies out that culture is a key predictor for the environmental behaviors based on a 27-country sample to predict proenvironment behaviors.

Regarding the perceived behavior control, other than the external factors like time, money, chance and etc., the individuals are known to perform a specific behavior when they are able to control over the opportunities and resources (Chen and Tung 2014). Perceived behavioral control is usually attached with the self efficacy which is the self belief in the own ability and the people with high self efficacy would rather think that they will make an impact on the issue as in the case of boycotting the luxury fashion apparel made in sweatshops (Phau et al. 2015). Price can be an other example for a control belief if there is perceived expensiveness regarding the price. Within this respect, if the price of an organic product is more expensive than a regular product, a consumer may prefer not to purchase the organic product (Maloney et al. 2014). Related with the control beliefs, some researchers stated that it can be taken by some people as a measure of effectiveness. Consumer effectiveness is described as the willingness of the concerned consumers to give back to the planet because of their belief that their individual actions will make a difference towards solving a problem (Laskova 2007).

As seen above, TPB is based on only the cognitive belief which are related mostly with thoughts and beliefs (Arvola et al. 2008) and therefore it is critisized because of insufficient consideration of affective aspects which are related with the emotions.

There are not much studies dealing with the integration of the emotions into the TPB however. Instead, some researchers preferred to analyse the influence of the awareness and knowledge on the beliefs and behaviors of the consumers.

In many studies, it was confirmed that the awareness and knowledge about the organic products has an impact on the attitude towards organic products (Maloney et al. 2014). These studies showed that the consumers don't know much about the meaning of certified organic products and labelling or organic products and even unaware of the benefits of organic products. Ellen et al. (1991) showed that consumers with higher concern for the environment and perceived consumer effectiveness have a higher level of perceived knowledge toward solving a problem. The effectiveness of consumer decision increases with the increase in awareness and knowledge about the environmentally friendly products (Stets and Biga 2003). Nonetheless, some studies showed that the awareness and knowledge about the organic products may also influences consumers' perception of expensiveness. For instance, Loureiro et al. (2002) analysed a negative perception regarding pricing of organic products and Wang (2007) noticed that the organic products are less appealing because of their price.

Last but not least, adapted from Robbins and Greenwald (1994), the sustainable attitude is proposed to have stages which were described as 'incorporative', 'impulsive', 'imperial', 'interpersonal' and 'institutional' respectively. The first stage corresponds to no awareness. In the second stage, the consumer is not aware but has capacity for knowledge, in the imperial stage, the consumers focus on the immediate effect but not future believing that they can't have impact. In the interpersonal stage, the consumers concern what others think and believes that individual has a little control or influence. And finally in the institutional stage, the consumers feel responsibility for their actions and has clear sense of beliefs (McNeill and Moore 2015).

4.2 Previous Studies with the Theory of Planned Behavior

Considering that it is possible to add some more constructs into the TPB model after the theory's current variables have been taken into account, TPB model was used either in its original form or in extended form in many studies. Below, the studies which applied TPB in the original and the extended form were summarized which were applied in different fields regarding sustainability and the textile and fashion industry specifically.

4.2.1 The Application of TPB in Different Fields

Tikir and Lehmann (2011) used TPB in order to find out the approach of the consumers toward public transport which became important because of the climate change. The author used a value based approach in combination with theory of planned behavior. Within this regard, they integrated 4 more construct called indi-vidualist, egalitarian (group oriented having the concern of in distribution of costs and benefits), fatalist (individual oriented favoring isolation) and hierarchical (supports current system and dislikes changes) The authors found out that based on predomi-nant value type within the cultural theory, the behavior via attitudes and norms can be perceived as positive or not by consumers.

Haustein and Hunecke (2007) used an extension of TPB to explain travel mode choice integrating the construct named as perceived mobility necessities. It was found that perceived mobility necessities moderated the relationship between public transportation attitude and intention.

Chen and Tung (2014) applied TPB model expanded with the environmental con-cern and perceived moral obligation to predict consumers' intention to visit green hotels. Perceived moral obligation was related with the responsibility of the individ-ual to perform a specific behavior morally in case of facing with ethical situation. Whereas the authors confirmed that all three main constructs of TPB were found to determine the purchase behavior regarding visiting the green hotels, the consumers' environmental concern was found to be positively related to his/her attitude toward visiting green hotels and the consumer's perceived moral obligation was found to have a positive impact on his/her intention to visit green hotels. The authors found out that perceived behavioral control was more important since the convenience and easy access could be important for selection of the hotels.

Shaw and Shiu (2002) assessed the ethical consumer decision making in fair trade grocery purchasing using theory of planned behavior modified with an addition of two construct which were ethical obligation and self identity. The ethical obligation was described with the expression an individuals internalized ethical rules which reflect personal beliefs about appropriate behavior whereas the self identity was described as the pertinent part of an individual's self that related to a particular behaviour. Nonetheless, the authors found contradictory results to previously made regression analysis and stated that the contribution of ethical obligation and self identity was low on the determination of the attitude.

Ozcaglar-Toulouse et al. (2006) further again investigated the consumer intention to purchase fair trade grocery products using theory of planned behavior by adding the same two constructs; self identity and ethical obligations. Dividing the sample into two parts which were entitled with the consumers, who purchase fair trade regularly and those who purchase rarely or never, the authors found out that the attitude and subjective norms were the key drivers of the intention for those who never or rarely purchase fair trade grocery products; whereas attitude and perceived behavioural control were the important factors influencing the intentions for those who regularly make such purchases. The authors added that the variables ethical obligation and self-identity added significant additional explanatory power for both subsamples. The three main antecedents of the intention in TPB were the attitudes which is the summed product of individuals' beliefs and their evaluation of those beliefs; subjective norms which was the summed product of individuals' beliefs that important others think they should or should not perform the behaviour in question, and their motivation to comply with those others and finally the perceived behavioural control which was related with the control beliefs concerning difficulties in conducting in the behaviour.

Yazdanpanah and Forouzani (2015) applied TPB in prediction of Iranian students to purchase organic food. The authors found out than only attitude was the significant predictor or the organic food purchase. The authors added that the integration of the moral norm and self identity to the model increased the explanatory power. The authors claimed that the young generation had higher awareness and more knowledge regarding the organic food. Adding that the higher order cognitive skills were deployed by interactions with the environment in the childhood. The authors indicated that the sample of the study was selected from students as the new generation was the consumers of the future.

Kim et al. (2012) attempted to find out if the eco-friendly consumer behavior was influenced by variations in social norms and by consumer's environmental concern using normative conduct theory. The author found out that the social norms were much more important for the consumers than the environmental concern which meant that observing the behaviors of others had great effect on purchase intention. Moreover, the marketing claim was found to be moderating the influence of injunctive norms and environmental concern on purchase intentions but the claim type was found not to differ the impact of descriptive norms on purchase intentions.

Considering that the self identification of the consumer to be making green consumerism showed a high correlation with the behavioral intention to buy organic foods, Irianto (2015) applied an extended TPB to analyse the variables affecting the consumer attitude to buy organic food in Indonesia. The variables included into the model were 'health consciousness', 'environmental consciousness', 'organic food price', 'attitude', 'subjective norm', 'intentions to purchase organic food' and gender. At the end of the study, the main determinants of the organic food purchase were found to be the health consciousness and environmental consciousness. Moreover, it was observed that gender made a difference on the attitudes, intentions, and finally the behaviors about purchasing organic food.

Arvola et al. (2008) applied TPB to predict purchase intentions of organic foods expanding the model with measures of affective and moral attitudes. The moral norm was added into the model in a way that it was connected with all the other constructs and it was described as a positive self enhancing feeling of "doing the right thing", instead of a negative feeling of obligation or guilt. The confirmation of the model was done by the authors but the effects of moral attitude measures were found to partially mediated via attitudes.

Hansen (2008) added values which could act as the guiding principles in one's living as the antecedents of the construct attitude in the application of TPB in the online grocery purchase. The values added within the model were given as openness to change, self-transcendence, conservation and self-enhancement. The author found that TPB was confirmed in predicting the purchase intention with these three antecedents for which the strength of the relationship was listed to decrease from attitudes to perceived behavioural control. On the other hand, the authors figured out that the attitudes which was the most powerful influencer of the intention is determined by the values.

Kim et al. (2013) analyzed the anticipated emotion in consumers' intention to select eco-friendly restaurants by using TPB, which was augmented by adding the construct regret. The authors stated that integrating the new construct was important since the cognitive based models were not so successful in measuring the feelings associated with the sources of information as indicated by Morris et al. (2002). According to the authors, humans were making their decisions in rational way in TPB model. Stating that regret could be defined as "a counterfactual emotion that was experienced in the present situation when imagining the results of a future outcome" (Bui 2009), it could be suggested as a powerful factor in motivating or directing the behaviour. The authors found out that the most influential factor for the original TPB model became the subjective norm which was followed with the attitude and perceived behavioural control. Whereas the sequence of the influential factors did not change for the extended TPB model with the factor regret, the construct regret took the third place in terms of strength on the intention coming before the perceived behavioural control.

4.2.2   The Application of TPB in Sustainable Consumption and Sustainable Fashion

Benefiting partially theory of reasoned action and theory of planned behavior, Mostafa (2007) investigated the effect of general values and specific values such as environmental concern and environmental knowledge on the green purchase attitude, intention and acceptance in the end. The author stated that the initial phase of green purchase process was the importance attached to environment and only the consumers for whom the environment was important could go through the purchasing process of information seeking and looking for green products. Moreover, knowing much more about the the eco-product by either getting information and having personal experience could develop specific beliefs which could further influence the choice. The author also related the beliefs with the islamic culture adding that the human beings acted as stewards for the God who created the nature. The author confirmed that the value-attitude-benefit approach was valid within the purchase intention of sustainable products underlying that the strength of the relationship between the intention and the actual purchase was not so high. On the other hand, regarding the values influencing the construct attitude, it was found that the perceieved environmental knowledge had higher impact on the attitude than the construct new environmental paradigm corresponding to the environmental concern.

Brosdahl and Carpenter (2010) investigated the environment friendly consumption behavior concentrating on the effects of the knowledge of the environmental impact and the concern for the environment. The authors claimed that the knowledge of the environmental impacts led to concern for the environment which then influenced the environmentally friendly consumption.

Paul et al. (2016) attempted to validate TPB in green product consumption applied the extended form of TPB including the environmental concern. The authors found out that the attitude and perceived behavioral control predicted purchase intention significantly whereas subjective norms did not have any influence. On the other hand, integrating the variable environmental concern into TPB model to analyse purchase intention yielded better model fit values.

Mancha and Yoder (2015) used environmental theory of planned behavior model in order to predict the green behavioral intentions working on a bi-national sample. Stating that the TPB model was ignoring the role of self identity and culture which was mainly shaping the identity, the authors included two variables as the antecedents of the attitude, social norms and perceived behavioral control. Adding the independent self construal and the interdependent self construal to the model, they developed a model with variance explained with a degree of 62.3%. It was found that independent self construal was much more influential on the attitude whereas the interdependent self construal was influential on the social norms and perceived behavior control.

Manchiraju and Sadachar (2014) investigated the influence of personal values on consumers' ethical fashion consumption behavior employing the Fritzsche model which put forward that the values were the predictors of the intentions. The authors benefited the 10 basic individual values which were (power; achievement; hedonism; stimulation; self-direction; universalism; benevolence; tradition; conformity; and security). Segmented these ten values into four clusters as suggested by Schwartz (1992, 1994), 'self-transcendence' values included benevolence and universalism; 'self-enhancement' values included hedonism, power, and achievement; 'conservation' included conformity, security, and tradition; and 'openness to change' included stimulation and self-direction. The authors found that the two groups of values which were influential on the behavioural intention were the self enhancement and openness. Moreover, the authors found out that women were more engaged in ethical fashion consumption than men. On the other hand, the authors revealed that whereas the consumption of organic fashion products were viewed as ethical consumption, fair trade fashion product consumption was not seen in this way by the consumers.

Hustvedt and Dickson (2009) investigated the likelihood of purchasing organic cotton apparel with an aim of explaining if the purchasing of organic cotton was a fashion trend and if there was a consumer base interested in purchasing organics cotton apparel because of the benefits of the organic agriculture. Believing that organic cotton usage had the common motivations with the purchasing organic foods, the authors stated that organic food consumption increased after the green consumerism movement which began in 1980s. They put forward the dilemma regarding the explanation of the motivation of the organic food purchasing to be the health reasons or the sustainability concerns. The authors claimed that whereas small portion of organic apparel consumers preferred to purchase because of health concerns, most of the consumers preferred to buy organic apparel because of the altruistic purchase motivations. The authors found out that the main motivation for the consumers to purchase organic apparel products was the belief about the beneficial outcomes of the purchase to themselves, the organic industry and the environment. Finally, the authors suggested that the marketing claims which stated that the consumers would help the environment indirectly by purchasing organic apparel would be helpful because of the altruistic characteristics of the consumers and the consumers being receptive to these kinds of marketing messages.

Kim and Karpova (2010) examined the consumer attitudes toward fashion counterfeits by applying TPB. The additional constructs were product appearance, past purchase behavior, value consciousness, and normative susceptibility, information susceptibility, value consciousness, integrity, status consumption and materialism. Among them the normative susceptibility was found to have negative relationship with the factor, attitude. The authors also studied different models analysing the additional links from subjective norms to attitudes and PBC. The most influential construct was found to be the subjective norms which was followed with the attitude and perceived behavioural control getting the lowest value in terms of strength. In the extended model, both additional paths were stated to increase the explanatory power of the TPB model.

Maloney et al. (2014) applied TPB on the consumer willingness to buy organic apparel. The researchers used the variables awareness, subjective norms, personal norms, attitude, perceived expensiveness, and perceived effectiveness as the factors influencing the purchase intention. The findings showed that the most significant factor was the attitude which was followed with the subjective norn on the construct intention. Personal values, on the other hand, were not found significantly influencing the intention construct. The construct attitude was found to be dependent on the perceived expensiveness and the awareness which was also influential on the perceived consumer effectiveness and perceived expensiveness. The construct perceived behavioural control did not take place within the model studied, and it was analysed within the additional construct perceived consumer effectiveness and perceived expectiveness which were found to be related with the construct attitude.

Phau et al. (2015) investigated the motivations for buying luxury fashion apparel made in sweatshops using TPB analysing the two cases which were intention not to purchase luxury branded fashion apparel made in sweatshop and the intention to pay more for the apparel not produced in the sweatshops. The authors described the social norms construct as the social pressure the person felt while establishing a certain behavior and perceived behavior control as the perception of the consumer about the ability to perform a behavior. The authors investigated that attitudes under the headings 'attitudes toward sweatshops', 'attitudes toward social consequences' and 'attitudes towards purchasing behavior' and social norms under the headings 'integrity', 'status consumption', 'information susceptibility', 'normative susceptibility'. The authors found out although the attitudes toward sweatshops did not influence neither the intention not to purchase luxury branded fashion nor the willingness to pay for more for ethical fashion, the other dimensions of the attitudes and perceived behavioral control were influential on these decisions. Regarding the social norms on the other hand, the integrity and status consumption were found to be significant on the intention not to purchase products produced in sweatshops; but only the integrity was found to be influential on the willingness to pay more.

Lavorata (2014) studied the influence of the retailers' commitment to sustainable development on the store image, consumer loyalty and consumer boycotts. She proposed a tool called RCSD (retailers' commitment to sustainable development as perceived by consumers) instead of the construct attitude in TPB and studied the impact of RCSD on store image, consumer loyalty and consumer boycotts. The authors developed the construct RCSD from the Triple Bottom Line which was a measurement tool for distinguishing three aspects of sustainable development which were economic (profit), social (people) and environmental (planet) (Lavorata 2014). Actually the authors confirmed that the two constructs taken directly from the TPB model, which was subjective norms and perceived behavioural control, were influential on the construct intention. The construct RCSD which was used instead of attitude was not found to be not influencing the purchase intention but found to be influential on the retailer's image.

5 The Empirical Study

5.1 Research Models and Hypotheses

In order to understand the consumers' behaviors, especially the younger consumers', towards sustainable fashion products, an empirical study was established in Turkey. Two models were employed within this study which were the original TPB model and extended TPB model. The original TPB model had the constructs the 'attitude', 'subjective norms' and 'perceived behavioral control' as the direct antecedents of the construct 'purchase intention'. The extended TPB model included the the additional constructs which were 'behavioral beliefs', 'normative beliefs' and 'control beliefs' respectively. These three construct were suggested to be the directly influencing the main constructs of TPB model in the order of being listed.

The additional constructs were included based on the studies established by (Han et al. 2010; Kang et al. 2013). Behavioral beliefs were described as the beliefs which could be developed after the behavior was decided to be applied. Normative beliefs were included corresponding to the moral values which were adopted by the close neighbourhood of the consumer. Control beliefs corresponded to the general opinions of the consumers about the sustainable fashion products which were mainly including the negative approaches.

The suggested constructs and the boundaries of the two research models were shown in Fig. 1 where the original TPB model included the continuous lines and the extended TPB model included both the continuous and the dashed lines.

Regarding the two research models, the following hypotheses were proposed to be validated

H1 Attitude toward the sustainable fashion products will have a significant effect on the intention to purchase the sustainable fashion products
H2 Subjective norms about the sustainable fashion products will have a significant effect on the intention to purchase the sustainable fashion products
H3 Perceived behavioral control about the sustainable fashion products will have a significant effect on the intention to purchase the sustainable fashion products
H4 Behavioral beliefs about the sustainable fashion products will have a significant effect on the attitude towards the sustainable fashion products

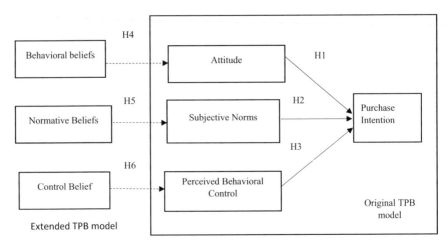

Fig. 1 The original TPB and the extended TPB research models

H5   Normative beliefs about the sustainable fashion products will have a significant
     effect on the subjective norms about the sustainable fashion products
H6   Control beliefs about the sustainable fashion products will have a significant
     effect on the perceived control beliefs about the sustainable fashion products.

5.2   Data Collection and Questionnaire

The data was provided from a survey conducted online. The questionnaire involved
sections for determining the socio-demographic characteristics of the participants
and the questions for the application of the original TPB and the extended TPB
models. This section included three items for 'purchase intention' (Ahn et al. 2007),
five items for 'attitude' (Han et al. 2010), three items for 'subjective norms' (Han
et al. 2010), three items for 'perceived behavioral control' (Han et al. 2010), five
items for 'behavioral beliefs' (4 items from Han et al. 2010; 1 item from attitude
Kang et al. 2013) three items for 'normative beliefs' (Han et al. 2010), and four
items for 'control beliefs' (Han et al. 2010). The items within the model section were
established using seven-point Likert scale which ranged from (1) disagree at all to
(7) completely agree.

5.3   Sample Profile

The questionnaire was conducted online within a time period of three-week period
of time. Among the 339 participants answered the questions, 28% of them were male

and 72% of them were female. 53.9, 36, 6 and 4.1% of the participants were belonged to four age groups which were given with the intervals '18–23', '24–29', '30–35' and '36 plus'. Regarding the education level of the participants, it could be stated that the majority of the participants corresponded to 42.7% which were 'still attending universities as undergraduate' whereas 36.3% of them were 'the under graduates'. These results were followed by three groups of the participants which were 'still attending graduate studies', 'the graduates' and 'the high school graduates' with the percentages 9.9, 6.1 and 5%. Since most of the participants were selected to be belonging to the younger age groups, the percentages of the participants became 41.7, 32.4, 14.6 and 11.3% for the 'low income class', 'low to medium class', 'medium class' and 'medium to high classes'.

5.4 Application Procedure

Data and the research models were analyzed by using IBM SPSS Version 25 and IBM AMOS (Analysis of Moment Structure) 21 programs. The univariate normality and multivariate normality of the data was tested by calculating the skewness and kurtosis of the data and by applying the Mardia's test of multivariate normality. No significant evidence for threat to univariate normality was found for both research models since all the data had the skewness values <3 and kurtosis values <10 as proposed by Kline (2005). Within the Mardia's test, the multivariate kurtosis got the value 131.905 and 319.203 for the original and extended TPB research models respectively. Since these values were smaller than 224 and 728 which were calculated based on the formula $p(p + 2)$, where p was taken as being to equal to the number of observed variables in the original TPB model and extended TPB model, the multivariate normality was assumed in this study (Raykov and Marcoulides 2008).

The scale variability was assessed by calculating Cronbach's alpha. Kaiser-Myer Ölkin (KMÖ) measure and Bartlett's test of sphericity were used to test sampling adequacy. The determinant of the correlation matrix was assessed to detect multi-collinearity. Exploratory factor analysis was used to determine factors and factor loadings. The models were then tested and modified using confirmatory factor analysis which was followed with several statistical validity tests and analyses such as reliability tests, convergent and discriminant validity tests. The hypotheses were verified and the strength of the relationships were assessed using structural equation modelling within the level of significance of 0.05.

For both the confirmatory factor analysis and structural equation modelling, model fit was assessed in term of the fit indexes and by comparing them with the recommended values which were given in Table 1.

Table 1 Model fit indices and the recommended values

Goodness of fit indexes	Values used for comparison	Reference
Chi-square/degree of freedom	<3	Kline (2005)
Goodness of fit index (GFI)	≥0.9	Hair et al. (2006)
Adjusted goodness of fit index (AGFI)	≥0.8	Chau (1996)
Normalized fit index (NFI)	≥0.9	Hair et al. (2006)
Comparative fit index (CFI)	≥0.9	Hair et al. (2006)
Root mean square residual (RMR)	≤0.08	(Hair et al. 2006; Brown and Cudeck 1992)
Root mean square of approximation (RMSEA)	≤0.08	Brown and Cudeck (1992)

5.5 Results

Before conducting the exploratory factor analysis, the data was confirmed to have the univariate and multivariate normality. Regarding the scale reliability, the Cronbach's alpha was found to be 0.885 and 0.887 for the original and extended TPB models which was well above the acceptable value of 0.70 (Nunnally 1978). The eligibility of the data for factor analysis was measured with the Kaiser-Meyer-Olkin and Bartlett's test of sphericity and it was found out that the sampling adequacies were found to be sufficiently large enough getting the values 0.872 and 0.910 for the original TPB model and extended TPB model respectively (Kaiser 1974). The results of the Bartlett's test of sphericity was also found significant for both research models. The determinant of the correlation matrices for both the original and the extended TPB model were checked and the multicollinearity was not detected since they got values greater than 0.00001 (Kinnear and Gray 1999).

The exploratory factor analysis was applied using principal axis factoring with varimax rotation for both the original TPB model and extended TPB model. The factor loadings of 14 items for original TPB model and 25 items for extended TPB model were given in Table 2. Since all the factor loading got values higher than 0.5, no items were deleted. It was seen that 14 items for original TPB model yielded four factors explaining 72.077% of the total variance. Nonetheless, it was seen that 25 items for the extended TPB model yielded six factors instead of seven factors as expected. The three items for the factor subjective norms and three items for the normative beliefs were loaded with the same factor. Within this respect, six factors in the extended TPB model were seen to explain 70.074% of the total variance.

The factor loading were also checked with confirmatory factor analysis. Some modifications were made in order to improve the model and the item PBC2 under the factor Perceived Behavioral control was eliminated from the model because of having negative variance within the error term of this item. The model fit values for

Table 2 The reliability and validity measures for the original and extended TPB models

Item	Original TPB model				Extended TPB model			
	Factor loading	CA	CR	AVE	Factor loading	CA	CR	AVE
Purchase intention								
PI1	0.739	0.876	0.879	0.708	0.615	0.876	0.879	0.707
PI2	0.857				0.774			
PI3	0.800				0.725			
Attitude								
ATT1	0.740	0.850	0.853	0.537	0.583	0.850	0.852	0.537
ATT2	0.668				0.587			
ATT3	0.625				0.651			
ATT4	0.820				0.764			
ATT5	0.746				0.689			
Perceived behavioral control								
PBC1	0.774	0.681	0.432	0.280	0.812	0.681	0.499	0.365
PBC2	0.824				0.826			
PBC3	0.718				0.628			
Subjective norms					Norms			
SN1	0.891	0.907	0.911	0.774	0.838	0.930	0.927	0.680
SN2	0.881				0.859			
SN3	0.765				0.689			
Normative beliefs								
NB1	NA	NA	NA	NA	0.797			
NB2					0.856			
NB3					0.809			
Behavioral beliefs								
BB1	NA	NA	NA	NA	0.718	0.858	0.864	0.563
BB2					0.822			
BB3					0.585			
BB4					0.746			
BB5					0.597			
Control beliefs								
CB1	NA	NA	NA	NA	0.697	0.725	0.744	0.426
CB2					0.738			
CB3					0.778			

both the original and extended TPB models before and after the modifications were given in Table 3.

In order to check the reliability and validity the Cronbach's alpha (CA) values, Composite reliability (CR) and the Average variance extracted (AVE) values were calculated after confirmatory factor analysis as seen in Table 2. The Cronbach alpha values became between 0.681 and 0.907 for the original TPB model and between 0.681 and 0.930 for the extended TPB model. The composite reliability values, which show the degree of the factor is represented by its items, were found to be between 0.432 and 0.879 for the original TPB model. Although the composite reliability of the factor 'Perceived behavioral control' getting the value 0.432 which was small according to Bagozzi and Yi (1988), the factor was not dropped from the analysis and no item was deleted from the factor because of having only 2 items in it. The composite reliability of the factors was found to be between 0.499 and 0.927 for the extended TPB model on the other hand. The convergent validity was also checked by calculating the average variance extracted values for both models. The average variance extracted values became between 0.280 and 0.708 for the orginal TPB model and between 0.365 and 0.707 for the extended TPB model. Since these values were approximate to or exceeded 0.5, the convergent validity of the factors were found to be satisfactory except the factor 'perceived behavioral control' (Bagozzi et al. 1991; Hair et al. 2006; de Holanda Coelho et al. 2016). Considering the average variance extracted values, again the convergent validity is not seemed to be satisfactorily for the construct 'perceived behavioural control'. But, since this construct was important for the TPB model, it was not eliminated for the further analysis. Finally, the divergent validity of the models which showed the extent, the items within each factor differ were assessed by comparing the square roots of average variance extracted values with the correlation between the factor and all the other factors (Teo et al. 2009). As the square roots of average variance extracted values were found to be higher than the correlations except for the construct 'perceived behavioural control', the discriminant validity of the models were satisfied.

The research models consisting of 4 factors and 6 factors respectively were tested by using structural equation modelling.

The results of the test revealed a reasonable model fit values by getting the values exceeding the required fit values which were given in Table 1. Moreover, in both the original and the extended models, the modification made by deleting one item from

Table 3 Model fit values of the original and extended TPB models

Model	x^2	Df	x^2/Df	GFI	AGFI	NFI	CFI	RMR	RMSEA
TPB model	149.545	69	2.167	0.941	0.910	0.944	0.969	0.087	0.059
TPB model with modifications	114.746	57	2.013	0.951	0.921	0.952	0.975	0.076	0.055
Extended TPB model	552.716	278	1.988	0.888	0.858	0.897	0.945	0.097	0.054
Extended TPB model with modifications	475.997	254	1.874	0.899	0.871	0.906	0.954	0.085	0.051

the construct 'perceived behavioral control' was needed. As seen from Table 3, this deletion improved the model fit values.

The hypotheses were tested and the estimates for the relationship strength were generated. The R^2 values were calculated for the independent values in both original TPB model and extended TPB model. Since the R^2 value became 0.563 for the independent variable 'intention' for the original TPB model and it is higher than 0.1, it was found to be appropriate and informative to examine the significance of the relations between the factors. Similarly, the R^2 values became 0.727 for the independent variable 'attitude' and 0.610 for the independent variable 'intention'. As these R^2 values were higher than 0.1, it was found that the significance of the relations associated with factors could be examined.

The relationships which had the p values lower than 0.05 were found to be statistically significant. Regarding the original TPB model, all the hypothesis H1, H2, and H3 were accepted. The strength of the relationships was estimated to be 0.644, 0.139 and 0.036 between the pairs 'attitude' and 'intention'; 'subjective norm' and 'intention'; and 'perceived behavioural control' and 'intention' respectively as seen in Fig. 2. Thus, the purchase intention for sustainable fashion products was much more influenced from the attitudes toward the sustainable fashion products. Although the strength of the relationship was only slightly higher, it could be stated that the secondarily influential factor on purchase intention was the subjective norms which were adopted because of the other people.

Regarding the extended TPB model, the hypotheses H1, H2, H3, H4, H5 were accepted but the hypothesis H6 was rejected since the p value was below 0.05. The strength of the relationships was estimated to be 0.694, 0.155, 0.031 for the relationship which were validated also in the original TPB model corresponding to the relations between the constructs 'attitude' and intention; 'subjective norms and normative beliefs' and the 'intention'; 'perceived control behavior' and 'intention'. Thus, the rank of the factors which were found to be influential on the purchase intention did not change for the extended TPB model. Regarding the additional factors,

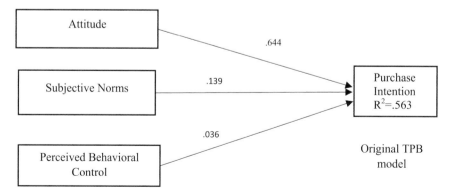

Fig. 2 The strength of the relationship for the original TPB model

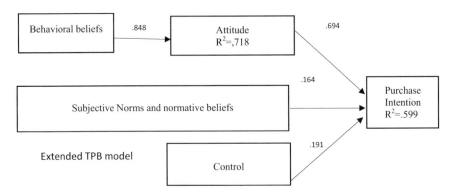

Fig. 3 The strengths of relation for the extended TPB model

it was found that the strength of the relationship between the factors 'behavioural beliefs' and the 'attitude' became 0.853. This meant that the attitudes toward sustainable fashion product is influenced from the behavioural beliefs of the consumers. The final presentation of the original and TPB and extended TPB model were given below with the strengths of the relationship which were found significant (Fig. 3).

Considering the findings, it can be stated that the major influential factor on the purchase intention of the sustainable fashion products is the attitude toward them which is followed by the subjective norms. Perceived behavioral control is found to be less influential factor on the purchase intention.

The same tendency is valid for the extended TPB model for which the addition of two construcfts which are 'behavioral beliefs' and 'normative beliefs' increased the explanatory power of the model. Behavioral beliefs is found to be major antecedent for the construct 'attitude' while the 'subjective norms' and 'normative belief' seem to create the same perception by the consumer as it is not possible to differentiate these two construct.

The least influential factor is found to be 'perceived behavioural control' for both models which meant that the perception of the consumers about the easiness of the behaviour to accomplish it does not have much impact on the construct 'intention'

Actually, the findings are quite similar with the previous findings of the researchers. The results confirm that TPB has a power to explain the purchase intention but it requires some more modification regarding the construct 'perceived behavioural control'. Although, it was not put forward by the researchers until so far, this situation can be related with the fact that the factor 'perceived behavioural control' can be much more influential on the behaviour instead of 'intention'. This means the construct 'control beliefs' does not determine the intention of the consumer before accomplishing the behaviour but just at the time of doing it. Thus, it can be proposed to integrate one more construct 'actual behavior' into the model to see if the same construct will have the same impact on that construct too.

Considering the previous findings given at the beginning of the chapter, the consumers actually have positive thoughts and beliefs about the sustainable fashion products but the reason why they avoid purchasing these products are probably related with the perception of them. The reason for the avoidance might be that the consumers avoid from these products because of thinking that they have poor quality, they are sold at high prices and most of them are not trendy or stylish. Since these points were out of scope of this study, it is impossible to assure exactly these claims but it can be recommended that the TPB model can be extended in a way that it also includes the actual behaviour where the construct 'perceived behavioural control' is also analysed as the mediating construct between purchase intention and actual behaviour. Last but not least one more construct is suggested to be implemented about the perception of sustainable fashion product including the characteristic other than being sustainable, ethical or ecological.

6 Conclusion

This chapter attempted to analyse the consumer behaviour towards sustainable fashion by pointing out the different aspects of sustainable fashion, summarizing the approach of consumers toward this concept and analyzing the consumer behaviour using a well known model.

The empirical study showed that the original TPB model and proposed TPB model were confirmed in analysing the consumer behaviour in parallel with the previous studies. Nonetheless, the effect of the construct 'perceived behavioural control' was found to be least influential control in both the original and proposed model where additional constructs were found to increase the explanatory power of the model. Being in parallel with the previous studies and based on the analysis regarding the approaches of the consumers towards sustainability concepts and issues, it was suggested that the control beliefs may not be influential on the purchase intention so much and they might be more influential on the actual behaviour. And it was suggested to analyse also the beliefs of the consumers about the sustainable fashion products such as quality, aesthetic and price.

References

Abrahamse, W., & Steg, L. (2009). How do socio-demographic and psychological factors relate to households' direct and indirect energy use and savings? *Journal of Economic Psychology, 30*(5), 711–720.

Aggeri, F., Pezet, E., Abrassart, C., & Acquier, A. (2005). *Organiser le développement durable: Expériences des entreprises pionnières et formation de règles d'action collective* (p. 278). Paris: Vuibert.

Ahn, T., Ryu, S., & Han, I. (2007). The impact of Web quality and playfulness on user acceptance of online retailing. *Information & Management, 44*(3), 263–275.

Ajzen, I. (1991). The theory of planned behavior. *Organizational Behavior and Human Decision Processes, 50*(2), 179–211.

Ajzen, I. (2002). Perceived behavioral control, self-efficacy, locus of control, and the theory of planned behavior. *Journal of Applied Social Psychology, 32*(4), 665–683.

Ajzen, I., & Fishbein, M. (1980). *Understanding attitudes and predicting social behaviour*.

Arnocky, S., Stroink, M., & DeCicco, T. (2007). Self-construal predicts environmental concern, cooperation, and conservation. *Journal of Environmental Psychology, 27*(4), 255–264.

Arvola, A., Vassallo, M., Dean, M., Lampila, P., Saba, A., Lähteenmäki, L., et al. (2008). Predicting intentions to purchase organic food: The role of affective and moral attitudes in the Theory of Planned Behaviour. *Appetite, 50*(2–3), 443–454.

Bagozzi, R. P., & Yi, Y. (1988). On the evaluation of structural equation models. *Journal of the Academy of Marketing Science, 16*(1), 74–94.

Bagozzi, R. P., Yi, Y., & Phillips, L. W. (1991). Assessing construct validity in organizational research. *Administrative science quarterly*, 421–458.

Bandura, A. (1986). *Social foundations of thought and action: A social cognitive theory*. Englewood Cliffs, NJ, US: Prentice-Hall Inc.

Banerjee, S., Gulas, C. S., & Iyer, E. (1995). Shades of green: A multidimensional analysis of environmental advertising. *Journal of Advertising, 24*(2), 21–31.

Birtwistle, G., & Moore, C. M. (2007). Fashion clothing—where does it all end up? *International Journal of Retail & Distribution Management, 35*(3), 210–216.

Bissonnette, M. M., & Contento, I. R. (2001). Adolescents' perspectives and food choice behaviors in terms of the environmental impacts of food production practices: Application of a psychosocial model. *Journal of nutrition education, 33*(2), 72–82.

Bonnes, M., & Bonaiuto, M. (2002). Environmental psychology: From spatial-physical environment to sustainable development. *Handbook of environmental psychology*, 28–54.

Bray, J., Johns, N., & Kilburn, D. (2011). An exploratory study into the factors impeding ethical consumption. *Journal of Business Ethics, 98*(4), 597–608.

Brosdahl, D. J., & Carpenter, J. M. (2010). Consumer knowledge of the environmental impacts of textile and apparel production, concern for the environment, and environmentally friendly consumption behavior. *Journal of textile and apparel, technology and management, 6*(4).

Browne, M. W., & Cudeck, R. (1992). Alternative ways of assessing model fit. *Sociological Methods & Research, 21*(2), 230–258.

Bui, M. (2009). *Consumer regret regulation: Examining the effects of anticipated regret on health-related decisions*. Arkansas, United States: University of Arkansas.

Carrigan, M., & Attalla, A. (2001). The myth of the ethical consumer—Do ethics matter in purchase behaviour? *Journal of consumer marketing, 18*(7), 560–578.

Casadesus-Masanell, R., Crooke, M., Reinhardt, F., & Vasishth, V. (2009). Households' willingness to pay for "green" goods: Evidence from Patagonia's introduction of organic cotton sportswear. *Journal of Economics & Management Strategy, 18*(1), 203–233.

Chan, K. (1998). Mass communication and pro-environmental behaviour: Waste recycling in Hong Kong. *Journal of Environmental Management, 52*(4), 317–325.

Chan, T. Y., & Wong, C. W. (2012). The consumption side of sustainable fashion supply chain: Understanding fashion consumer eco-fashion consumption decision. *Journal of Fashion Marketing and Management: An International Journal, 16*(2), 193–215.

Chao, Y. L. (2012). Predicting people's environmental behaviour: Theory of planned behaviour and model of responsible environmental behaviour. *Environmental Education Research, 18*(4), 437–461.

Chau, P. Y. (1996). An empirical assessment of a modified technology acceptance model. *Journal of management information systems, 13*(2), 185–204.

Chen, A., & Peng, N. (2012). Green hotel knowledge and tourists' staying behavior. *Annals of Tourism Research, 39*(4), 2211–2216.

Chen, M. F., & Tung, P. J. (2014). Developing an extended theory of planned behavior model to predict consumers' intention to visit green hotels. *International Journal of Hospitality Management, 36*, 221–230.

Clark, H. (2008). SLOW + FASHION—An Oxymoron—or a Promise for the Future …? *Fashion Theory, 12*(4), 427–446.

Corraliza, J. A., & Berenguer, J. (2000). Environmental values, beliefs, and actions: A situational approach. *Environment and Behavior, 32*(6), 832–848.

Cowe, R., & Williams, S. (2000). *Who are the ethical consumers?*. Co-operative Bank: Ethical Consumerism Report.

Davis, G., Phillips, P. S., Read, A. D., & Iida, Y. (2006). Demonstrating the need for the development of internal research capacity: Understanding recycling participation using the Theory of Planned Behaviour in West Oxfordshire, UK. *Resources, Conservation and Recycling, 46*(2), 115–127.

de Holanda Coelho, G. L., Hanel, P. H., Medeiros Cavalcanti, T., Teixeira Rezende, A., & Veloso Gouveia, V. (2016). Brief Resilience Scale: Testing its factorial structure and invariance in Brazil. *Universitas Psychologica, 15*(2), 397–408.

Dean, M., Raats, M. M., & Shepherd, R. (2012). The role of self-identity, past behavior, and their interaction in predicting intention to purchase fresh and processed organic food. *Journal of Applied Social Psychology, 42*(3), 669–688.

Domeisen, N. (2006). When ethics meet fashion. In *International Trade Forum* (Vol. 3, No. 2).

Ellen, P. S., Wiener, J. L., & Cobb-Walgren, C. (1991). The role of perceived consumer effectiveness in motivating environmentally conscious behaviors. *Journal of Public Policy & Marketing*, 102–117.

Erdem, O., Ben Oumlil, A., & Tuncalp, S. (1999). Consumer values and the importance of store attributes. *International Journal of Retail & Distribution Management, 27*(4), 137–144.

Ferdous, A. S. (2010). Applying the theory of planned behavior to explain marketing managers' perspectives on sustainable marketing. *Journal of International Consumer Marketing, 22*(4), 313–325.

Fielding, K. S., McDonald, R., & Louis, W. R. (2008). Theory of planned behaviour, identity and intentions to engage in environmental activism. *Journal of environmental psychology, 28*(4), 318–326.

Fishbein, M., & Ajzen, I. (1975). *Belief, attitude, intention and behavior: An introduction to theory and research*.

Fletcher, K. (2010). Slow fashion: An invitation for systems change. *Fashion Practice, 2*(2), 259–265.

Floyd, L. A., Xu, F., Atkins, R., & Caldwell, C. (2013). Ethical outcomes and business ethics: Toward improving business ethics education. *Journal of Business Ethics, 117*(4), 753–776.

Hair, J. F., Black, W. C., Babin, B. J., Anderson, R. E., & Tatham, R. L. (2006). *Multivariate data analysis* (6th ed.). Englewood Cliffs, N.J.: Prentice Hill.

Han, H., Hsu, L. T. J., & Lee, J. S. (2009). Empirical investigation of the roles of attitudes toward green behaviors, overall image, gender, and age in hotel customers' eco-friendly decision-making process. *International Journal of Hospitality Management, 28*(4), 519–528.

Han, H., Hsu, L. T. J., Lee, J. S., & Sheu, C. (2011). Are lodging customers ready to go green? An examination of attitudes, demographics, and eco-friendly intentions. *International Journal of Hospitality Management, 30*(2), 345–355.

Han, H., Hsu, L. T. J., & Sheu, C. (2010). Application of the theory of planned behavior to green hotel choice: Testing the effect of environmental friendly activities. *Tourism Management, 31*(3), 325–334.

Hansen, T. (2008). Consumer values, the theory of planned behaviour and online grocery shopping. *International Journal of Consumer Studies, 32*(2), 128–137.

Harland, P., Staats, H., & Wilke, H. A. (1999). Explaining proenvironmental intention and behavior by personal norms and the theory of planned behavior. *Journal of Applied Social Psychology, 29*(12), 2505–2528.

Haustein, S., & Hunecke, M. (2007). Reduced use of environmentally friendly modes of transportation caused by perceived mobility necessities: An extension of the theory of planned behavior. *Journal of Applied Social Psychology, 37*(8), 1856–1883.

Heath, Y., & Gifford, R. (2002). Extending the theory of planned behavior: Predicting the use of public transportation. *Journal of Applied Social Psychology, 32*(10), 2154–2189.

Hermans, C. M., Schaefer, A. D., & Haytko, D. (2007). A cross-national examination of the dimensionality of the consumer susceptibility to interpersonal influence scale. *International Journal of Business Research, 7*(5), 186–191.

Herr, P. M., Kardes, F. R., & Kim, J. (1991). Effects of word-of-mouth and product-attribute information on persuasion: An accessibility-diagnosticity perspective. *Journal of Consumer Research, 17*(4), 454–462.

Horng, J. S., Su, C. S., & So, S. I. A. (2013, July). Segmenting food festival visitors: Applying the theory of planned behavior and lifestyle. In *Journal of Convention & Event Tourism* (Vol. 14, No. 3, pp. 193–216). Taylor & Francis Group.

Hustvedt, G., & Bernard, J. C. (2008). Consumer willingness to pay for sustainable apparel: The influence of labelling for fibre origin and production methods. *International Journal of Consumer Studies, 32*(5), 491–498.

Hustvedt, G., & Dickson, M. A. (2009). Consumer likelihood of purchasing organic cotton apparel: Influence of attitudes and self-identity. *Journal of Fashion Marketing and Management: An International Journal, 13*(1), 49–65.

Irianto, H. (2015). *Consumers' attitude and intention towards organic food purchase: An extension of theory of planned behavior in gender perspective.*

Jang, J., Ko, E., Chun, E., & Lee, E. (2012). A study of a social content model for sustainable development in the fast fashion industry. *Journal of Global Fashion Marketing, 3*(2), 61–70.

Joergens, C. (2006). Ethical fashion: Myth or future trend? *Journal of Fashion Marketing and Management: An International Journal, 10*(3), 360–371.

Johnson, C. Y., Bowker, J. M., & Cordell, H. K. (2004). Ethnic variation in environmental belief and behavior: An examination of the new ecological paradigm in a social psychological context. *Environment and Behavior, 36*(2), 157–186.

Kaiser, H. F. (1974). An index of factorial simplicity. *Psychometrika, 39*(1), 31–36.

Kalafatis, S. P., Pollard, M., East, R., & Tsogas, M. H. (1999). Green marketing and Ajzen's theory of planned behaviour: a cross-market examination. *Journal of Consumer Marketing, 16*(5), 441–460.

Kals, E., Schumacher, D., & Montada, L. (1999). Emotional affinity toward nature as a motivational basis to protect nature. *Environment and Behavior, 31*(2), 178–202.

Kang, J., Liu, C., & Kim, S. H. (2013). Environmentally sustainable textile and apparel consumption: The role of consumer knowledge, perceived consumer effectiveness and perceived personal relevance. *International Journal of Consumer Studies, 37*(4), 442–452.

Keh, H. T., & Xie, Y. (2009). Corporate reputation and customer behavioral intentions: The roles of trust, identification and commitment. *Industrial Marketing Management, 38*(7), 732–742.

Kidwell, B., & Jewell, R. D. (2003). An examination of perceived behavioral control: Internal and external influences on intention. *Psychology & Marketing, 20*(7), 625–642.

Kim, Y., & Han, H. (2010). Intention to pay conventional-hotel prices at a green hotel—A modification of the theory of planned behavior. *Journal of Sustainable Tourism, 18*(8), 997–1014.

Kim, H., & Karpova, E. (2010). Consumer attitudes toward fashion counterfeits: Application of the theory of planned behavior. *Clothing and Textiles Research Journal, 28*(2), 79–94.

Kim, H., Lee, E. J., & Hur, W. M. (2012). The normative social influence on eco-friendly consumer behavior: The moderating effect of environmental marketing claims. *Clothing and Textiles Research Journal, 30*(1), 4–18.

Kim, Y. J., Njite, D., & Hancer, M. (2013). Anticipated emotion in consumers' intentions to select eco-friendly restaurants: Augmenting the theory of planned behavior. *International Journal of Hospitality Management, 34*, 255–262.

Kim, J., & Park, J. (2005). A consumer shopping channel extension model: Attitude shift toward the online store. *Journal of Fashion Marketing and Management: An International Journal, 9*(1), 106–121.

Kinnear, P. R., & Gray, C. D. (1999). *SPSS for Windows made simple.* Taylor & Francis.

Kline, R. B. (2005). *Principles and practice of structural equation modeling* (2nd ed.). New York: Guilford press.

Kotchen, M. J., & Reiling, S. D. (2000). Environmental attitudes, motivations, and contingent valuation of nonuse values: A case study involving endangered species. *Ecological Economics, 32*(1), 93–107.

Laeequddin, M., & Sardana, G. D. (2010). What breaks trust in customer supplier relationship? *Management Decision, 48*(3), 353–365.

Lai, K. H., Wong, C. W. Y., & Cheng, T. C. E. (2012). Ecological modernization forces and performance implications of green logistics management. *Technological Forecasting and Social Change, 79*(4), 766–770.

Laskova, A. (2007). Perceived consumer effectiveness and environmental concerns. *Proceedings of the Asia Pacific Management Conference, Australia, 13,* 206–209.

Lavorata, L. (2014). Influence of retailers' commitment to sustainable development on store image, consumer loyalty and consumer boycotts: Proposal for a model using the theory of planned behavior. *Journal of Retailing and Consumer Services, 21*(6), 1021–1027.

Liao, C., Chen, J. L., & Yen, D. C. (2007). Theory of planning behavior (TPB) and customer satisfaction in the continued use of e-service: An integrated model. *Computers in Human Behavior, 23*(6), 2804–2822.

Loureiro, M. L., McCluskey, J. J., & Mittelhammer, R. C. (2002). Will consumers pay a premium for eco-labeled apples? *Journal of Consumer Affairs, 36*(2), 203–219.

Maloney, J., Lee, M. Y., Jackson, V., & Miller-Spillman, K. A. (2014). Consumer willingness to purchase organic products: Application of the theory of planned behavior. *Journal of Global Fashion Marketing, 5*(4), 308–321.

Mancha, R. M., & Yoder, C. Y. (2015). Cultural antecedents of green behavioral intent: An environmental theory of planned behavior. *Journal of Environmental Psychology, 43,* 145–154.

Manchiraju, S., & Sadachar, A. (2014). Personal values and ethical fashion consumption. *Journal of Fashion Marketing and Management, 18*(3), 357–374.

Markus, H. R., & Kitayama, S. (1991). Culture and the self: Implications for cognition, emotion, and motivation. *Psychological Review, 98*(2), 224.

McGoldrick, P. J., & Freestone, O. M. (2008). Ethical product premiums: Antecedents and extent of consumers' willingness to pay. *The International Review of Retail, Distribution and Consumer Research, 18*(2), 185–201.

McNeill, L., & Moore, R. (2015). Sustainable fashion consumption and the fast fashion conundrum: Fashionable consumers and attitudes to sustainability in clothing choice. *International Journal of Consumer Studies, 39*(3), 212–222.

Miller, C. (1992). Levi's, Esprit spin new cotton into eco-friendly clothes. *Marketing News, 26,* 11–12.

Minton, A. P., & Rose, R. L. (1997). The effects of environmental concern on environmentally friendly consumer behavior: An exploratory study. *Journal of Business Research, 40*(1), 37–48.

Morris, J. D., Woo, C., Geason, J. A., & Kim, J. (2002). The power of affect: Predicting intention. *Journal of Advertising Research, 42*(3), 7–17.

Mostafa, M. M. (2007). A hierarchical analysis of the green consciousness of the Egyptian consumer. *Psychology & Marketing, 24*(5), 445–473.

Niinimäki, K. (2010). Eco-clothing, consumer identity and ideology. *Sustainable Development, 18*(3), 150–162.

Norwegian Ministry of Environment. (1994, January 19–20). *Symposium: Sustainable consumption.* Oslo, Norway: Ministry of the Environment.

Nunnally, J. C. (1978). *Psychometric theory.* New York: McGraw-Hill.

Oreg, S., & Katz-Gerro, T. (2006). Predicting proenvironmental behavior cross-nationally: Values, the theory of planned behavior, and value-belief-norm theory. *Environment and Behavior, 38*(4), 462–483.

Ozcaglar-Toulouse, N., Shiu, E., & Shaw, D. (2006). In search of fair trade: ethical consumer decision making in France. *International Journal of Consumer Studies, 30*(5), 502–514.

Paul, J., Modi, A., & Patel, J. (2016). Predicting green product consumption using theory of planned behavior and reasoned action. *Journal of Retailing and Consumer Services, 29,* 123–134.

Pelling, E. L., & White, K. M. (2009). The theory of planned behavior applied to young people's use of social networking web sites. *CyberPsychology & Behavior, 12*(6), 755–759.

Phau, I., & Ong, D. (2007). An investigation of the effects of environmental claims in promotional messages for clothing brands. *Marketing Intelligence & Planning, 25*(7), 772–788.

Phau, I., Teah, M., & Chuah, J. (2015). Consumer attitudes towards luxury fashion apparel made in sweatshops. *Journal of Fashion Marketing and Management, 19*(2), 169–187.

Pookulangara, S., Hawley, J., & Xiao, G. (2011). Explaining consumers' channel-switching behavior using the theory of planned behavior. *Journal of Retailing and Consumer Services, 18*(4), 311–321.

Raykov, T., & Marcoulides, G. A. (2008). *an introduction to applied multivariate analysis.* New York: Taylor & Francis.

Robbins, J. G., & Greenwald, R. (1994). Environmental attitudes conceptualized through developmental theory: A qualitative analysis. *Journal of Social Issues, 50*(3), 29–47.

Roberts, J. A. (1996). Green consumers in the 1990s: Profile and implications for advertising. *Journal of Business Research, 36*(3), 217–231.

Rogers, R. W. (1983). Cognitive and psychological processes in fear appeals and attitude change: A revised theory of protection motivation. *Social Psychophysiology: A Sourcebook*, 153–176.

Rosenstock, I. M. (1974). The health belief model and preventive health behavior. *Health Education Monographs, 2*(4), 354–386.

Schaefer, A. (2005). Some considerations regarding the ecological sustainability of marketing systems. *Electronic Journal of Radical Organisation Theory, 9*(1), 40.

Schwartz, S. H. (1977). Normative influences on altruism. In *Advances in experimental social psychology* (Vol. 10, pp. 221–279). Academic Press.

Schwartz, S. H. (1992). Universals in the content and structure of values: Theoretical advances and empirical tests in 20 countries. In *Advances in experimental social psychology* (Vol. 25, pp. 1–65). Academic Press.

Schwartz, S. H. (1994). Are there universal aspects in the structure and contents of human values? *Journal of Social Issues, 50*(4), 19–45.

Schwartz, S. H., & Howard, J. A. (1984). Internalized values as motivators of altruism. In *Development and maintenance of prosocial behavior* (pp. 229–255). Boston, MA: Springer.

Shamdasani, P., Chon-Lin, G. O., & Richmond, D. (1993). Exploring green consumers in an oriental culture: Role of personal and marketing mix factors. *ACR North American Advances.*

Shaw, D., & Riach, K. (2011). Embracing ethical fields: constructing consumption in the margins. *European Journal of Marketing, 45*(7/8), 1051–1067.

Shaw, D., & Shiu, E. (2002). An assessment of ethical obligation and self-identity in ethical consumer decision-making: A structural equation modelling approach. *International Journal of Consumer Studies, 26*(4), 286–293.

Shen, B., Wang, Y., Lo, C. K., & Shum, M. (2012). The impact of ethical fashion on consumer purchase behavior. *Journal of Fashion Marketing and Management: An International Journal, 16*(2), 234–245.

Sheth, J. N., Sethia, N. K., & Srinivas, S. (2011). Mindful consumption: a customer-centric approach to sustainability. *Journal of the Academy of Marketing Science, 39*(1), 21–39.

Sparks, P., & Shepherd, R. (1992). Self-identity and the theory of planned behavior: Assesing the role of identification with "green consumerism". *Social psychology quarterly*, 388–399.

Stern, P. C., Dietz, T., Abel, T., Guagnano, G. A., & Kalof, L. (1999). A value-belief-norm theory of support for social movements: The case of environmentalism. *Human Ecology Review*, 81–97.

Stets, J. E., & Biga, C. F. (2003). Bringing identity theory into environmental sociology. *Sociological Theory, 21*(4), 398–423.

Stryker, S. (1968). Identity salience and role performance: The relevance of symbolic interaction theory for family research. *Journal of Marriage and the Family*, 558–564.

Tang, J. H., & Farn, C. K. (2005). The effect of interpersonal influence on softlifting intention and behaviour. *Journal of Business Ethics, 56*(2), 149–161.

Tarkiainen, A., & Sundqvist, S. (2005). Subjective norms, attitudes and intentions of Finnish consumers in buying organic food. *British Food Journal, 107*(11), 808–822.

Teo, T., Lee, C. B., Chai, C. S., & Wong, S. L. (2009). Assessing the intention to use technology among pre-service teachers in Singapore and Malaysia: A multigroup invariance analysis of the Technology Acceptance Model (TAM). *Computers & Education, 53*(3), 1000–1009.

Thøgersen, J., & Ölander, F. (2002). Human values and the emergence of a sustainable consumption pattern: A panel study. *Journal of Economic Psychology, 23*(5), 605–630.

Tikir, A., & Lehmann, B. (2011). Climate change, theory of planned behavior and values: a structural equation model with mediation analysis. *Climatic Change, 104*(2), 389–402.

Vermeir, I., & Verbeke, W. (2008). Sustainable food consumption among young adults in Belgium: Theory of planned behaviour and the role of confidence and values. *Ecological Economics, 64*(3), 542–553.

Walker, A. E., Grimshaw, J. M., & Armstrong, E. M. (2001). Salient beliefs and intentions to prescribe antibiotics for patients with a sore throat. *British Journal of Health Psychology, 6*(4), 347–360.

Wang, P. (2007). *Consumer behavior and willingness to pay for orgnaic [ie organic] products.*

Watson, M. C., Johnston, M., Entwistle, V., Lee, A. J., Bond, C. M., & Fielding, S. (2014). Using the theory of planned behaviour to develop targets for interventions to enhance patient communication during pharmacy consultations for non-prescription medicines. *International Journal of Pharmacy Practice, 22*(6), 386–396.

Whitmarsh, L., & O'Neill, S. (2010). Green identity, green living? The role of pro-environmental self-identity in determining consistency across diverse pro-environmental behaviours. *Journal of Environmental Psychology, 30*(3), 305–314.

Wiseman, M., & Bogner, F. X. (2003). A higher-order model of ecological values and its relationship to personality. *Personality and Individual Differences, 34*(5), 783–794.

World Commission on Environment and Development (WCED). (1987). *Our common future: The Brundtland report.* New York: Oxford University Press.

Yazdanpanah, M., & Forouzani, M. (2015). Application of the Theory of Planned Behaviour to predict Iranian students' intention to purchase organic food. *Journal of Cleaner Production, 107,* 342–352.

Zemore, S. E., & Ajzen, I. (2014). Predicting substance abuse treatment completion using a new scale based on the theory of planned behavior. *Journal of Substance Abuse Treatment, 46*(2), 174–182.

Zhou, Y., Thøgersen, J., Ruan, Y., & Huang, G. (2013). The moderating role of human values in planned behavior: The case of Chinese consumers' intention to buy organic food. *Journal of Consumer Marketing, 30*(4), 335–344.

Sustainable Sports Fashion and Consumption

Elaine Fung and Rong Liu

Abstract The trend of promoting sustainable and healthy lifestyle is fast growing. Consumers paying more rational attention upon the goods they purchased have become a global norm. Meanwhile, the introduction of fast fashion incurred huge amount of fashion waste, driving consumers' more heed towards environmental issues caused by the fashion industry. To gain more market share, fashion retailers realized that capturing the consumers' demands in a sustainable fashion manner would be highly important to generate higher revenue in a long run. However, fashion products with sustainable features usually cost relatively higher, which results in a higher market price. In this chapter, we will explore several key factors upon fashion consumer behaviors. Then, we will focus on the growing compression sportswear, especially seamless sports fashion market to analyze how sportswear companies design and produce sustainable sportswear by using new fashion technology to sustain business. Following this, we will further discuss how the sportswear companies adopt sustainability concept throughout the whole functional sportswear design and development process.

Keywords Sustainability · Consumer behavior · Sports fashion
Seamless knitting · Compression sportswear

1 Global Fashion Market

Due to the financial constraints, people in the past may not afford to purchase new clothes other than necessities. They can only purchase new clothes when needs are arisen and problems (defects) are recognized. For instance, they buy new clothes when old clothes are outworn or when the weather changes.

E. Fung · R. Liu (✉)
Institute of Textiles and Clothing, the Hong Kong Polytechnic University,
Hung Hom, Hong Kong
e-mail: rong.liu@polyu.edu.hk

© Springer Nature Singapore Pte Ltd. 2019
S. S. Muthu (ed.), *Consumer Behaviour and Sustainable Fashion Consumption*,
Textile Science and Clothing Technology,
https://doi.org/10.1007/978-981-13-1265-6_2

Fig. 1 Number of employees in apparel manufacturing, textiles manufacturing and textiles & clothing industry from year 1990 to 2015. *Reference* Catalyst Corporate Finance (2014)

When times goes by, as the discretionary income become higher and the living standard is improved, people can purchase any clothes they want with higher frequencies. Fashion products are no longer necessities but become products to satisfy people's unlimited wants which may over their actual needs. The needs can be the physiological needs, social needs, esteem needs or self-actualization needs, etc. This implies that the purchasing motives have changed, and it varies from consumer to consumer.

Traditionally, the fashion is mainly divided by 4 seasons annually, which are spring, summer, fall and winter (some of the retailers like American Rag, have 2 additional seasons, i.e., Holiday and Trans). To generate sales, retailers target the endless desire of satisfaction fulfillment and develop products to attract customers' attention. In the 20th century, concept of fast fashion was introduced. Fashion retailers like Zara and H&M are the pioneers. They introduced 16–52 seasons (monthly/weekly season) annually, which brings newness to consumers in a very timely manner and boots the sales dramatically. To catch up with the ever-changing fashion trends, consumer visit the stores and check the newly launch items from the fast fashion stores. Unlike the traditional fashion retailers, the fast fashion retailers tense to have small lots for each style (Martínez et al. 2015), and items commonly are sold out soon, which gives the consumer a perception that they need to buy quickly, otherwise they would fail to purchase the latest in-store items and cannot keep up the trend.

The growth in fashion industry resulted from the introduction of fast fashion was shown by the increase of employees in the apparel manufacturing, textiles manufacturing and textiles and clothing industries (Fig. 1). It was estimated that the global apparel market valued at 3 trillion US dollars which accounts for 2% of the world gross domestic product (GDP) (Catalyst Corporate Finance 2014).

As consumers' disposable income increase, they start seeking for fashion apparel products to meet their lifestyle and to satisfy their psycho-physiological needs. In recent years, with aids of information technologies, consumers commonly share their daily life via internet, e.g. Facebook, Instagram, Snapchat, wei-bo and WeChat etc. Refreshing their status via those platforms become one of their routines, which meanwhile draws an increasing attention on health and well-being lifestyle. Playing sports has become one of most popular types of lifelong healthy fashion to benefit to the physical body and help mind stay fresh.

2   Global Sportswear Market

With rising of casual sports, the consumption of sportswear is growing globally. The statistical analyses indicated that middle class living in the urban environment increase the demand in lifestyle sports (Rannikko et al. 2016). It is reported that the global retail sales in 2013 is US$263 billion with 15.3% contributed by sportswear sales and the forecast shows that growth in sportswear sales will keep the increase trend. The project growth of sportswear sales in 2017 was 7.5%, which is higher than the growth in apparel sales. The expected growth of sportswear in sales in North America is USD 9.1 billion, USD 16 billion in Eastern Europe, USD 28 billion in Western Europe, US 56 billion in Asia Pacific, USD 15 billion in Middle East/Africa, 8 billion in Latin America and 4 billion in Australia (Catalyst Corporate Finance 2014).

The significant growth in sportswear market supports the increase in sportswear business. For example, the latest global brands ranking shows that among all the fashion brands (Luxury, fast fashion, mass market retailers, etc.), sportswear brands Nike positioned in the 1st ranking in sales by 8% increase in 2017. Moreover, as the market of sportswear increase dramatically, many new comers enter the sports market, e.g. Patagonia and Lululemon, which focus more on outdoor clothing and yoga sportswear. To now, they have become significant players in sportswear market.

2.1   Compression Sportswear Market and Product

As the sportswear market competition become keener, compression sportswear has become one of emerging sportswear categories to draw retailers to invest. Professional athletes engaging diverse sports activities have increasingly selected compression sportswear in their daily training and international competitions. Liu and Little established a 5Ps contextual model to optimize athletic wear comfort in sports (Fig. 2). They considered that compression sportswear can be examined by the athlete regarding her/his perception of five categories: physical, psychological, physiological, psychophysical, and psychophysiological properties. Multi-relationships exist between the athlete, athletic wear, immediate body space and sports environment and culture. The 5Ps model explores the mechanisms of action of body-clothing-sports environment system from a comprehensive view, to guide optimizing functional

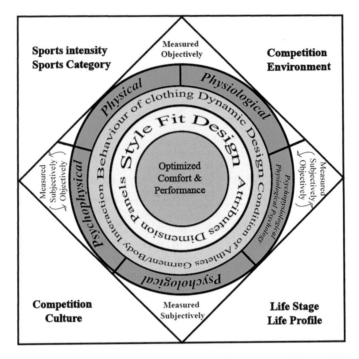

Fig. 2 5Ps Contextual model to optimize athletic wear comfort in sports. *Reference* The 5Ps model to optimize compression athletic wear comfort in sports (Liu and Little, Journal of Fiber Bioengineering and Informatics, 2009)

design of compression sportswear in practice for enhanced sports achievement and comfort (Liu and Little 2009).

Due to the wide spread of lifestyle sports (Rannikko et al. 2016) (e.g. jogging, running, skate boarding, yoga and roller etc.), people gradually treat them as a daily routine and have increasing demands for their functional properties. They are willing to pay more to purchase a set of professional athlete's compression sportswear with advanced functional performances (e.g. temperature regulation, moisture transfer and reduction of potential injury) and delightful wearing experience (comfort tactile and pressure perception).

The high profit margin of compression sportswear attracts sportswear brands to expand their business correspondingly. Sportswear retailers intend to balance the aesthetic design and product functionality to capture more market share and drive the brand recognition. They placed high investment in designing and developing compression sportswear with special functionalities towards specific types of sports. Table 1 and Fig. 3 show the examples of compression sportswear products provided by the sportswear brands in the current market.

Table 1 Characteristics of compression sportswear

No.	Style		Major materials[a]	Functional properties	Structure
1	Full Body Garment	LST	80% polyester/20% spandex	Promote natural range of motion via raglan sleeves Thumbholes keep sleeves in place	Plan (warp knits)
		FLB	80% polyester/20% spandex	Tight fit design provides confident and supported	Plan (warp knits)
2	Full Body Garment	LST	92% polyester/8% spandex	Promote natural body movement by shaping shoulders and arms	Plan (warp knits)
		FLB	92% polyester/9% spandex	Fast-drying fabric wicks sweat quickly.	Plan (warp knits)
3	Full Body Garment	LST	54% nylon/34% polyester/12% spandex	Facilitate natural range of motion by stretch fabric Easy layering via body-skimming design	Plan (warp knits)
		FLB	54% nylon/34% polyester/12% spandex	Keep tights in place by elastic waistband	Plan (warp knits)
4	Full Body Garment	LST	88% recycled polyester/12% spandex	Keep wearer dry and comfortable	Plan (warp knits)
		FLB	88% recycled polyester/12% spandex	Keep wearer dry and comfortable	Plan (warp knits)

LST Long Sleeves top; *FLB* Full length bottom; [a]at main body

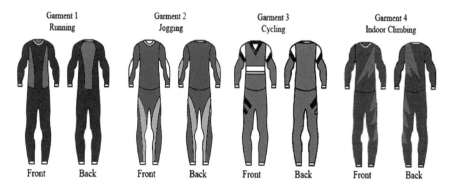

Fig. 3 Compression sportswear applied in different sports items

According to the Allied Market Research,[1] the global compression wear and shapewear market will be over 5.5 million US by 2022. Meanwhile, due to the high government investment upon the sports and changes in lifestyle fashion, the growth in Asia Pacific will be the highest among all global regions.

In year 2000 Sydney Olympic game, "fastskin" swimsuits were introduced. Almost 85% of the gold medal winners wear the fastskin swimsuit, which brought a huge attention to the compression sportswear. To date, compression garments for sportswear and leisure applications have become widely available providing improved comfort, fit and muscle support. Table 2 shows different types of form-fitted Athletic wear with respect to Olympic Sports and their physical demands correspondingly.

Based on the above findings, retailers need to have thorough understanding upon the consumers' requirements and may even need to provide massive information to satisfy consumers' desires to product knowledge. Different sports and leisure sports activities have different requirements on functionalities of compression sportswear. A number of lab and field studies have demonstrated that compression sports could have several advantages to enhance sports performances (Hayes and Venkatraman 2016), e.g. enhancing blood circulation (Yeager 2010), reducing recurrence of injury (Reynolds 2016), aiding recovery (Piras and Gatta 2017; Engel and Sperlich 2016), providing muscle support, reducing muscle soreness, and enhancing lactic acid removal, etc. (Fig. 4). The details are as follows.

- *Enhancing blood circulation* The mechanism of compression sportswear is analogous to compression therapy treatment which is to apply a certain amount of pressure to the body to facilitate venous return. For example, the highest pressure is exerted at the ankle or wrist in compression suits and decreasing gradually up towards the heart to aid the blood circulation.
- *Reducing the recurrence of muscular injury* The support of compression garment reduces the vibration or enhances the stabilization of skeletal muscles in movement, which brings down the possibility of injuries caused by intensive or the long-term exercises (Lovell et al. 2011; Marqués-Jiménez et al. 2017; Kraemer et al. 1998).
- *Aiding recovery* As the compression sportswear helps enhance blood circulation, the blood can further bring the metabolic waste away from muscles to speed up recovery. Researchers found that wearing compression sportswear for 24 h following exercise can reduce the soreness perceived by the athletes, then assist them to return normal training timely (Lovell et al. 2011; Berry and Mcmurray 1987; Davies et al. 2009; Duffield et al. 2010).
- *Reducing muscle soreness* During exercise, body parts may get in touch to the ground or the sports instruments, which would generate strikes to the body. The strikes will then cause vibration to the muscles and induce undesired shake, which may damage the muscles and lead to post-exercise soreness. The pressure exerted

---

[1]Compression Wear and Shapewear Market Forecast 2014–2022; https://www.alliedmarketresearch.com/compression-wear-shapewear-markt.

Table 2 Different types of athletic wear with respect to Olympic sports and its physical demand correspondingly

Olympic sports	Athletic category	Physical demands	Athletic wear
Out-door			
Track and Field Athletics	Field events, long jump, shot put	Speed, endurance, power	Form Fitted/Loose
Canoeing	Kayak, straight racing	Endurance, speed	Form fitted
Equestrian	A horseback rider	Precision, skill	Semi-form fitted
Football (Soccer)	Ball game, team sport	Speed, skill, precision, endurance	Semi-form fitted
Tennis	Racquet sport	Stamina, skill, speed	Semi-form fitted
Triathlon	Multi-sport event	Endurance, skill, power, speed	Form-fitted
Baseball	Ball game, team sport	Strength, teamwork, speed, precision/power	Form-fitted
Beach Volleyball	Team sport	Agility, speed, precision	Form-fitted
Archery	Shooting sports	Precision, skill	Semi-form fitted
Cycling Mountain Bike	Cycling	Endurance, power, speed	Form-fitted
Cycling Road	Cycling	Endurance, power, speed	Form-fitted
Goft	Stroke sports	skill, precision	Semi-form fitted
In-door			
Gymnastics	Competitive exercise	Agility, power, coordination, stabilization	Form fitted
Judo	Martial art and combat sport	Skill, power	Loose
Table Tennis	Racquet sport	Skill, speed	Semi-form fitted
Wrestling	Martial arts	Power, stabilization, skill	Loose
Weight Lifting	Weight training	Balance, flexibility and consistency, strength	Form-fitted
Badminton	Racquet sport	Aerobic stamina, agility, strength, speed and precision	Semi-form fitted

(continued)

Table 2 (continued)

Olympic sports	Athletic category	Physical demands	Athletic wear
Basketball	Ball game, team sport	Strength, teamwork, speed, precision/power	Loose
Boxing	Combat sport	Skill, power, speed, defense	Loose
Cycling Track	cycling	Endurance, power, speed	Form-fitted
Fencing	Competitive swordsmanship	Speed, precision, skill	Semi-form fitted
Handball	Team sport	Speed, stamina, skill, precision	Semi-form fitted
Shooting	Shooting sports	Precision, skill	Semi-form fitted
Hockey	Team sport	Speed, stamina, skill, precision	Loose
Taekwondo	Combat sport	Skill, strength, power	Loose
Water sports			
Swimming	Competitive water sport	Speed, endurance, power	Form-fitted
Diving	Competitive water sport	Speed, skill, precision	Form-fitted
Rowing	Competitive water sport	Speed, skill, stamina	Form-fitted
Sailing	Competitive boating	skill, precision	Form-fitted
water polo	Team sports	Speed, stamina, skill, precision	Form-fitted

Reference Liu et al. (2012)

by compression sportswear may hold the muscles in place and protect them from undesired vibration thus to reduce soreness (Marqués-Jiménez et al. 2017).

- *Enhanced lactic acid removal* To sustain intensive physical activity of the athletes, muscle needs to fill with oxygen. When the oxygen in the muscle is deficient, lactic acid will be generated, which allow the athletes to feel tired and lose energy. To re-oxygenate the muscle, the heart pumps the oxygenated blood to the muscles, and brings back the deoxygenated blood together with the lactic acid from the muscles. The gradient pressures produced by the compression sportswear may help push the blood flow from the distal body gradually towards the heart, thus enhancing the lactic acid removal (Berry and Mcmurray 1987).

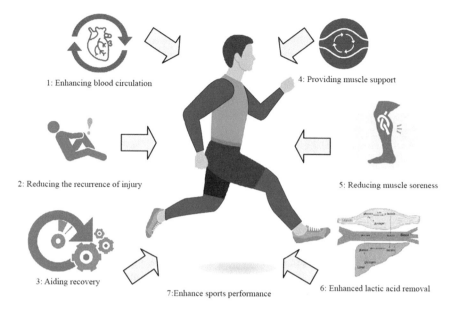

1: Enhancing blood circulation

4: Providing muscle support

2: Reducing the recurrence of injury

5: Reducing muscle soreness

3: Aiding recovery

7: Enhance sports performance

6: Enhanced lactic acid removal

Fig. 4 An example of functional requirements of compression sportswear

2.2 Sustainable Sports Fashion

As fashion apparel are no longer durable goods, the more fashion apparel being consumed, the more the waste it implies. Fashion waste not only includes the waste of the end product which the consumer disposes after use, it also includes the waste generated during the whole product development process, starting from planning, design, and manufacture to launch. Every year tons of wastes are generated by the fashion industry, and a growing awareness lead to the idea of sustainable fashion. Fast fashion brand H&M is now targeting 100% recycle materials for all its products. In year 2014, H&M collected 13,000 tons of wastes textile instore which is equivalent to 65 million T-shirts and the number is increasing and expected to be 100,000 tons in year 2020. To show that H&M is truly commitment to environmental protection and sustainability, they even host a panel discussion in London and discuss their green-friendly attitude towards fashion apparel design and development (Szmydke and Marfil 2015).

As people place more attention to the environmental friendly issues, more and more studies relating to sustainable fashion are being conducted and reported. We found 289 out of 372 SCI (Science Citation Index) journals (78%) in recent 10 years by using "sustainable fashion" as the keyword in searching (Fig. 5).

The sustainable fashion trend is also captured by the outerwear and sportswear brands. North Face incorporated various approaches to enhance their levels of sustainable supply chain management (Shen et al. 2017); Patagonia is using recycle or organic materials in new product design; Sportswear leading brand Nike has

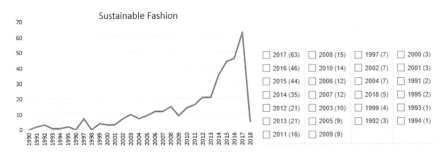

Fig. 5 Sustainable Fashion related SCI journals

transitioned sustainable business and innovation function into the company's core innovation function in 2013.

Sustainable development has board meanings. It concerns how we can sustain the planet in a long run (Gallopín and Raskin 2002) and how we can make good use of natural materials and improve our life without affecting the materials which our next generation can enjoy. That is, sustainable development framed as the need not only for improving the quality of life of all people now but also into long term future (WECD 1987). To assist companies to achieve sustainability in a more systematic manner, researchers attempted to bring the concept of life cycle sustainability assessment (Hannouf and Assefa 2017) and supply chain operation management theories into consideration (Shen et al. 2017; Shen 2014; de Brito et al. 2008; Karaosman et al. 2016).

One of the most popularly used theoretical model is the triple bottom line model (TBL) (Park and Kim 2016; Hiller Connell and Kozar 2017). It includes 3 core areas, i.e., economic, environmental and social. Sportswear retailers like Columbia have taken the 3 core elements of TBL model (i.e., economic, environmental and social sustainability) into product development process (Fig. 6).

- *Triple bottom line—environment*

From the environmental point of view, retailers select the environmental friendly features of the product as their preferences in business plan. For example, design for

Fig. 6 Core elements of
triple bottom line model

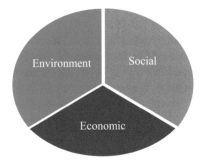

environment (DFE); design for recyclability (DFR); design for reusability(DFRU); design for re-manufacturability (DFRM), design for disassembly(DFD), and design for maintainability/serviceability (DFM/DFS). Sportswear companies like Kusaga Athletic, their core business concept comes from DFE. Their investigation indicates that processing 1 t-shirt takes 3000 L of water. The water pollution is a huge issue when more than 2 billion sport t-shirts were sold every year. Inspired by this, "Greenest Tee" was introduced. This new process only consumes 1% of water compared to the traditional processing. The fabric they used is called as **ECOLITE®**, which is made by blends of biodegradable natural fibers to fulfill the concept of DFE.[2]

- *Triple bottom line—social*

From the social point of view, sportswear companies need to consider how they can benefit the society throughout the whole product development process. For example, company needs to commit to the community and ensure all the labor they hired are well treated and all the materials they used are purchased by fair trade. Retailer like Columbia pays high attention on sustainable social responsibility, e.g., they input efforts in improving the social and ethical performance, and work closely with vendors who share the same ethical value, and provide a safe working condition to the workers. To educate the consumers upon their sustainable concepts, they attached swing tickets on the sportswear to show which kinds of sustainable perspectives they used in the products. Consumers who cares about shopping ethically will support the growth of business and thus further encourage the retailer to do better[3] (Fig. 7).

- *Triple bottom line—economic*

In the sense of economic sustainability, at the plant level, it has been operationalized as production or manufacturing costs (Cruz and Wakolbinger 2008). Traditionally, companies will only focus on the economic growth from their business, e.g., adopting non-environmental friendly materials or production process to bring down the cost and maximize the profit, but which would be harmful to the society. To achieve sustainable economic status, sustainable factors need to be considered, where the introduced extra cost may bring up the company cost and may decrease the revenue. However, it implies that part of the revenue will go to the participants who contribute to sustainable development and make the participants to be able to continue the business and contribute to achieve a sustainable planet. The growth of this company will generate a win-win situation for both economic growth and environment protection. For example, Nike launched a brand-new style of trainer through applying advanced flat-bed knitting technology in footwear industry. They work with Stoll (leading flat knitting machine supplier) and developed the Nike Flyknit running shoes in year 2012. According to their sustainable business report 2014–15, the flyknit technology saved 2 million pounds of waist in 2012, which encouraged Nike to continue refining its performance with 28 models across 6 categories in year 2015. The sustainable

---

[2]Kusaga; Athletic; https://www.kusagaathletic.com/pages/the-greenest-tee-shirt-on-the-planet.

[3]Outdoor Clothing, Outerwear and Accessories; https://www.columbia.com/About-Us_Corporate-Responsibility.html.

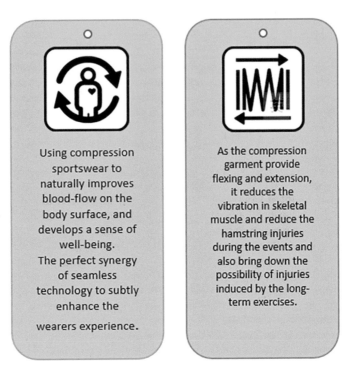

Fig. 7 Example of swing tickets attached onto the sportswear

new products not only contribute to growth of Nike's sales and revenue, but also boosting partner's innovation in new sustainable fabrication technologies. This can keep the sustainable business rolling to benefit the both economy and environment.[4]

2.3 Sustainable Seamless Compression Sportswear

Although fashion companies expect the business opportunity on promoting sustainable image, it is not an easy task to achieve sustainable fashion, especially when it comes to functional sportswear. On one hand, the products need to fulfill basic requirement for wearing comfort and functional requirements (e.g. the 5P's); on the other hand, it needs to take all the sustainable features into consideration during the product development process. Integrating the triple bottom line model with the 5Ps model, we can develop a sustainable sportswear model upon seamless compression sportswear development (Fig. 8).

The planning for sustainable seamless compression sportswear is fundamentally same as the traditional planning for normal garments. It starts with product planning.

---

[4]Adidas United States; http://www.adidas.com/us/parley.

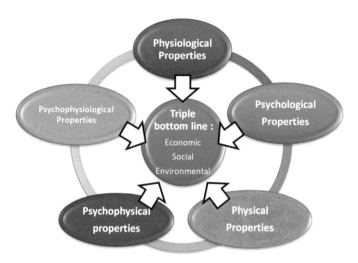

Fig. 8 Integrated model of 5P's and triple bottom line

Sportswear company plan what items they want to sale. The planner will work with sales team and marketing team upon the product details and features that they are planning to sale and promote. They will then work with the designers about the total product planning.

One of the main considerations in planning is that what kinds of manufacturing technologies they are going to adopt. For compression sportswear, it can be made by traditional cut & sew method or using seamless knitting techniques. The both have totally different requirements towards materials selection, as well as design and manufacturing lead-time. To achieve functional requirements and development sustainability of compression sportswear, seamless knitting manufacturing could be one of recommended choices.

- *Sustainable design and development process*

Traditional design and development process of fashion apparel induces a lot of waste. After getting information from planning teams on the items, designers will draft different designs and have the manufacturers make actual prototypes for review and approval to catch up the coming seasons. Normally each new design may need only several prototypes. Applying the traditional sampling machines to develop prototypes will consume numerous energy and water in multi-step manufacturing process (e.g. spreading, dying, cut out fabrics, assembly, etc.). Manufactures may only select few of them to produce at end. The prototypes those are not selected will become industrial waste for disposal. However, in the modern manufacturing, the process of prototype making can be largely shorten. Applying Computerized Aided Design and Computerized Aided Manufacturing (CAD/CAM) techniques, designers can generate the look of the finished sportswear in 2D (Fig. 9) or 3D effects in computer,

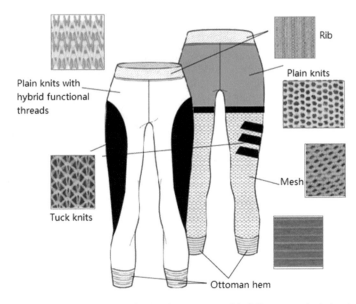

Plain knits with hybrid functional threads

Tuck knits

Rib

Plain knits

Mesh

Ottoman hem

Fig. 9 An example of design effect of knitted sportswear with different panel stitches in 2D view

and the fitting effects can be amended via software and displayed without making the actual garment.

When the design has been finalized, a mock up made by raw white materials without dying can be produced for details adjustment. This process not only save a lot of time and make the design more efficient, but also it decreases the material waste and pollutant generated.

- *Sustainable production*

Fashion apparel manufacturing can be mainly divided into cut & sew, and sew-free methods (e.g. bonding and fully fashion). For woven material based cut and sew manufacturing, garments are made by joining different panels to form 3-dimensional shape to fit the body. During manufacturing, patterns of garment panels are generated and placed onto plies of woven fabrics which is spreaded on the spreading table. As the patterns are in irregular shapes, there exist gaps between each panel patterns (i.e., the marker making efficiency <100%). Those "gap fabrics" will become waste materials, and the dyes and treatment that implies in these "gap fabrics" will become waste too. The case of cut and sew knitting manufacturing is similar to woven one. The knitted fabrics are commonly made by circular or warp knitting units. The tubular fabrics will be cut open and spread on the spreading table, so that garment panels can be cut from the spreader fabrics. Compared with it, sweater is more materials saving, which mainly apply v-bed knitting machines in fabrication. The panels of the garment can be knitted in shape by fully fashion technology to save materials. However, extra time is demanded to link the panels together loop by loop. If the garment is in fine gauge, e.g., 16gg (16 needle loops in one inch), it takes hours to

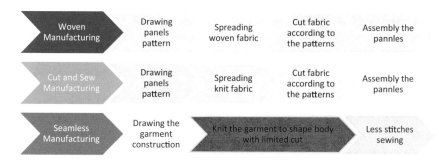

Fig. 10 A comparison on woven, Cut and Sew, and seamless knitting manufacturing

finish linking one garment, which is much slower than cut and sew manufacturing. Apart from the hourly wages of the workers, it is also a high burden to the factory on energy usage, as more energy is needed to keep good working environment (e.g. lightings, air conditioning, and power up the linking machines, etc.).

In this sense, seamless knitting manufacturing should be the one of the most sustainable types of apparel manufacturing among all. Seamless compression sportswear can be knitted by circular seamless knitting unit (e.g. SANTONI seamless series, Italy). Unlike cut and sew manufacturing, seamless knitting does not need to open up the tubular, instead, to use machines with different diameters, gauges and needle sizes to achieve the required body dimension or shape. For example, if we use a needle cylinder with diameter of $15''$, a tubular knitting body with M (middle) size without side seams can be produced. And also, by using the needle selection system, different stitches structures and patterns can be knitted in one-piece of seamless knitting body to realize specific aesthetic, physical-mechanical or thermo-moisture properties (Fig. 10).

Retailers usually produce more than the forecast which is called safety stock, as they think any "out of stock" means loss of sales and profit. However, as the demand is uncertain, when the actual demand is lower than expected, left over stock will be incurred. If retailer cannot sell the stock even discount and promotion are made, the stock will become dead stock and finally become waste. Seamless knitting adopts piece-by-piece working principle, which requires less processing procedures, thus providing possibility to avoid the above mentioned problem. For example, the finished garment parts can be knitted directly from the seamless knitting machines, and limited sewing is required to link the panels together. Taking a seamless knitted tank-top for an example, few stitches are only needed to sew two short shoulder seams and neck line of the whole garment piece. The assembly time and production materials are largely saved. Retailers can quickly response to market demands and place orders closer to the time of sales or even during the sales season, resulting in less dead stock and fashion waste.

3 Consumer Altitudes and Purchase Decision

Based on the above sections, we further explore the opportunity of sustainable sports fashion market and illustrate how seamless compression sportswear can achieve the target of sustainability. Consumers' altitudes and attention are critically important to generate sales and consumption, which would be an initial driver for functional sportswear market (Muzikante and Reņģe 2011). Rowley suggested five main roles in the decision-making process (Rowley 1997), which are, (1) users, who actually use the product or service; (2) influencers, particularly those with previous experience of the service; (3) deciders, the actual decision makers in the use or purchase decision; (4) approvers, who finally authorize the decision within an organization; and (5) buyers, those with the formal authority to buy and act as gatekeepers for purchasing (Hanna and Wozniak 2001).

In the case of seamless compression sportswear, the user is the key to make the consumption decision. As compression sportswear is a kind of personalized sportswear item (next to skin), which has higher requirements on tactile comfort, fitting and functionality. The users can be classified into 4 levels, i.e., the professional athletes, sports elites, sports trainers and sport lovers (Fig. 11). Most of the users are the sports lovers. They treat sports as a daily routine, so comfort and fit of sportswear could be their main requirements in use. Sports trainers and elites tend to have professional training, they more seek to sportswear which can enhance their exercise and training performances. Among all the sports players, the professional athletes (the top level) have stricter demands to sportswear quality and function. For example, they pursue sportswear to stabilize muscle groups in intensive exercise and to speed up their recovery after competition. If the retailers have better understandings on customer's attitudes and their specific demands in different user levels, they will deliver right sportswear products to satisfy the target market, thus promoting the sales. During this process, the attitudes of consumers are one of key factors affecting decision making process.

3.1 Consumer Attitude

Attitudes are learned or acquired rather than inborn. They are formed because of personal experience, reasoning or information, and the communicated experience of others (Fishbein and Ajzen 1975). A person's overall attitude towards an object is seen to be a function of the strength of each belief that a person holds about various aspects of the object, and the evaluation he gives to each belief as it relates to the object (Loudon and Bitta 1984).

In the retailing point of view, evaluating the consumers' attitude is very beneficial and useful for decision making. As the consumer attitudes represent how they react to the advertisements, brands, products, stores and so on. These attitudes are reflected through their consumption patterns, because consumers satisfy their needs

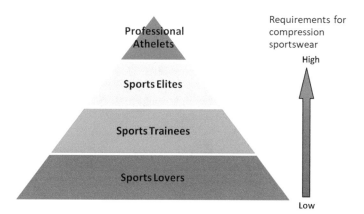

Fig. 11 4 levels of sport participants in market

Fig. 12 Consumer attitude functions

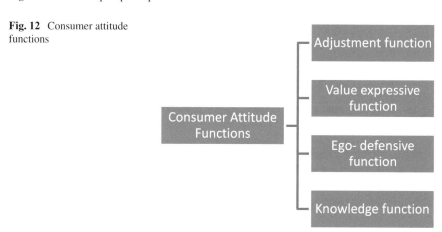

and wants commonly through consuming products and services. In a nut shell, attitude can be viewed as explainers and predictors of a wide variety of marketplace behavior. Providing sufficient product information, sportswear brands can link the consumers' beliefs and attitudes with the sustainable sportswear's features to facilitate consumers' action of consumptions.

There are four major attitude functions, (1) the adjustment function, (2) the value expressive function, (3) the ego-defensive function, and (4) the knowledge function. These functions provide a general direction for the sportswear companies to predict purchasing behavior of their target customers (Loudon and Bitta 1984; Hanna and Wozniak 2001) (Fig. 12).

- *The adjustment functions*

The adjustment function directs people toward pleasurable or rewarding objects and away from unpleasant or undesirable ones. It serves as the concept of utility maxi-

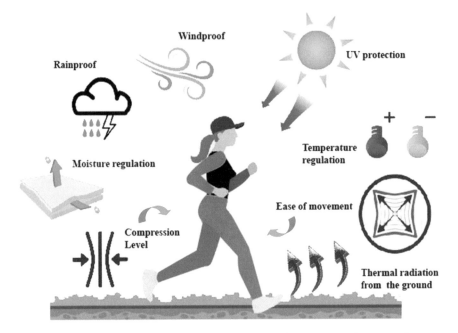

Fig. 13 Design a sportswear according to the requirements of sports type and sports environment

mization. As people tend to maximize happiness and minimize punishment, the attitudes of consumers depend to a large degree on their perceptions of what is needed to be satisfied and what is punished. Learning from experiences, consumers may have different attitudes toward different objects and the attitude may change in different occasions. Katz (1960) has noted that two conditions are necessary to change an attitude that serves the adaptive function: (1) the attitude and activities related to it no longer provide satisfaction they once did, (2) the individual's level of aspiration has been raised. For example, sports lovers who have relatively lower requirements in sportswear may have positive attitudes towards economic priced compression sportswear. While the sports elites could prefer professional sports suits to commensurate her status. The high performance sportswear would be taken more interactive elements among body, environment and clothing into account during product design and development process (Fig. 13).

- *The value-expressive functions*

Attitudes are expression of value when they reflect the central value that an individual hold and prefers. The value-expressive function enables the expression of person's value and self to others, this helps to project individual's real life. Through product consumption, consumers can express their values to the external world (Molkenthin 2012). Therefore, a way to success for the sportswear retailers is that, identifying the vale that the sports players want to express about themselves, and based on it, to design compression sportswear to satisfy self-expression. For example, a yoga sport trainer wants to buy professional functional sportswear to show her level of proficiency among her peer group. Lululemon (a well-known professional yoga sportswear brand) could be one of her choices. Comparatively, a sport lover may prefer to choose Stella McCartney to show her fashion sense.

- *The ego-defensive functions*

Attitude is formed to protect the ego or self-image by emphasizing his or her place in the social world. It serves to protect the person from threats or internal feelings of treads. Thus, individuals who believe that others dislike them may project on their own unconscious feeling of hostility toward others. Attitude that serves the function of protecting the individual from unpleasant truths about themselves are virtually difficult to change by simple conversations and communications (Loudon and Bitta 1984; Hanna and Wozniak 2001). This attitude function is difficult for the sportswear brand to deal with as well. To promote the product and service (fitness center), the retailer sometimes need to convince the targeted consumers that they are physically unpleasant (for example, the body figure is not fit enough). If individuals really have the feeling of unpleasant, they may buy the product or service to compensate the negative feeling. But in most cases, the ego-defensive attitudes are unpredictable. In the case of sustainable compression sportswear, the sportswear brands need to make the target customers to be aware that they are not professional enough if they do not buy from their brand. Sportswear brand like 2XU indicated that their products were equipped with the most advanced compression technology to sustainably achieve quality compression level in durable use.

- *The knowledge functions*

Attitudes serve as a knowledge function when they help us organize and simplify the complex and inexplicable world in which we live. Katz (1960) points out that, individuals not only require beliefs in the interest of satisfying various needs, they also seek knowledge to give meaning to what would otherwise be an unorganized chaotic universe. People need standards and frames of reference for understanding their world, and attitudes help supply such standards. Another view of the functional issue is that attitude serves one basic function, that of social adaptation (Lynn 1984). This means that like other social cognitions like value, knowledge function is very useful. As this function helps people to acquire useful information and estimate information. This helps to direct our behavior efficiently which enables us to adapt to the environment in the best way. As our aforementioned, sustainability is a wide concept, the attitudes towards the purchase of sportswear are complicated. By integrating the 5P's model with the triple bottom line model, sportswear companies can illustrate the details of 5P's with respect to their products features to the consumers; meanwhile they can show the company's sustainable mission along with the product description, so that sportswear consumers can link the information received together and generate consumption and sales.

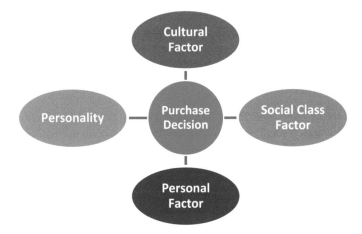

Fig. 14 Factors affecting consumer purchase decision

3.2 *Factors Affecting Sportswear Purchase Decision*

Understanding consumer's consumption attitude is crucial for successful business. In Sect. 3, we discussed how sportswear brands can evaluate their target customers by investigating consumer's attitude functions. So how to allow consumers to generate positive attitudes towards their products and further transform the positive attitudes to actual consumption? Below, we explore the factors that affect the consumer purchase decision (Fig. 14).

- *Cultural factor*

Culture is a society's distinctive and learned mode of living, interacting, and responding to environmental stimuli which is shared between members. Standards of acceptable behavior are passed along from one generation to the next in a continuation of the cultural heritage. Cultural learning occurs in various way, including living in the family environment, formal schooling, observing and imitating the behavior of others to serve as role models, and interact with other people within the society (Yolanda et al. 2017; Yurchisin and Johnson 2010).

   Apart from the society, family, peer groups, as well as the mass media and advertising (Hanna and Wozniak 2001), all these cultural backgrounds help us develop our own cultural norms and values (Arnould et al. 2004). In addition, culture and subculture affect consumer behavior in at least two ways, i.e., indirect and direct. Indirectly, they provide the background against most of our daily tasks and activities. Directly, they bear on the types of product available, purchase and consumed within the societies (Harrell 1986). For example, Hong Kong is an international city, adapting the free trade policy, brands have relatively free space to run their business. Almost all kinds of products from different countries globally can be found in Hong Kong market. Thus, there is no barrier for the athletes to consume any kind of sports apparels from different brands and countries. Consumers can get in touch with huge amount of sports apparel information and advertisement from those universal brands. With the higher level of education and mass information culture in Hong Kong, people tend to search for information before determining to buy the products. They also pay more attention in quality and brand reputation of the products to be purchased.

- *Social class factor*

Social class refers to the overall rank assigned to large group of people, according to the value held by the societies. Marketers are keenly interested in the distribution of wealth across the market place, because this phenomenon largely demarcates which group of people exhibits the greatest purchasing power (Yurchisin and Johnson 2010; Li et al. 2010). Nowadays, the annual income of major metropolitan is increasing. People concern more about lifestyle and are more willing to spend money on buying sportswear products. To show their social class, people attempt to choose professional functional sportswear. Thus, sportswear brands are willing invest more in functional product development, e.g. Lululemon, a yoga-inspired athletic apparel company, developed striking mesh design in sports tight to meet functional features in recent season. Even for those people who are less wealthy, they would like to buy sportswear with professional look but less functionality, this also bring new business opportunity to the retailers, e.g. some fast fashion brands, H&M, forever 21, expand their business by opening new diffusion lines of active sportswear.

- *Personal factor*

Consumer's purchasing behavior is usually affected by the people surrounding them. Due to reference groups and family members shape people's beliefs, attitudes and behavior, people whom we interact with like salesperson can also influence people's buying behavior. These interactions between two or more persons frequently influence the ideas, feelings or conduct of one or more participants in the conversations, especially for the women's market. This phenomenon is known as personal influence.

Personal influence refers to any change, whether deliberate or inadvertent, in an individual's beliefs, attitude or behavior that occurs as the consequences of interpersonal communication. Word of mouth (WOM) is one of the channels, through which personal influence can occur. WOM is a person-to-person communication between a non-commercial sources and receiver. In reality, WOM recommendations of product services, brands and stores can be transmitted in person, over the phone, through e-mail, or via internet. In the case of sustainable sportswear apparel, the opinion of the reference groups, family members and salesperson are highly important.

Personal influence can be verbal, visual or both. Consumer observe what other people are doing, wearing, or using, and may ask them what an item is, where it was purchased and how much it costs. This practice of information-sharing or exchanges can be one way or mutual. For the compression sportswear, as it is a kind of functional wear, consumer will seek for others opinions, especially those pier group who have used the product before, or through comments from public media, e.g. Instagram and Facebook, to inference functional performance of compression sportswear to be purchased (Lee 2003; Samli 2013; Jung and Kim 2016).

- *Personality*

Each person has their own personality. Personality is the sum of an individual's inner psychological attributes. It makes individuals what they are, distinguishes them from every person, demonstrates their mode of adjustment to life's circumstances, and produces their uniqueness and stable pattern of responding to environmental stimuli (Hanna and Wozniak 2001).

According to Freud personality theory (Burger 1993, Phares and Chaplin 1997), personality is a result of an interaction between the three components of personality: the id, the superego and the ego. The id is an unconscious reservoir of human instincts that seeks pleasure and demands immediate satisfaction. The superego is a largely unconscious repository of social, moral and ethical codes that restricts how far a person can go to acquire goals and steers instinctive drives into acceptable avenues. The ego is the conscious control center that mediates the id's desire encounter the ego's logic and the superego's prohibitions (Hanna and Wozniak 2001).

One view of personality assumes that humans possess traits that determine their personality. In the other words, if an individual behave in a sociable, adaptable, achievement-oriented, or independent manner, it is because that person possesses the corresponding traits.

In addition, Neo–Freudian's personality theory is a view that social variables rather than biological instincts and sexual drivers underlie personality formation. This theory directs marketers' attention to the social character of consumption. Like many campaigns emphasize social relation and human interaction. Sportswear brand Nike, introduces a new selling concept "Nike iD" which allows customers to design their own trainers, and customers can even sew their initial on the back of the trainer. This kind of customization allows consumers to show their own personality.[5]

4   Sportswear Consumer Decision Process (CDP) Model

To have an effective launching of sustainable sportswear, retailers need to have thorough understanding on consumer decision process. The Consumer Decision Process (CDP) Model represents roadmap of a consumer in a consumption action. It indicates the thought of consumer before, during and after the consumption. The CDP model provides the best guide of further study and evaluation. It is a useful method to analyze and explain why consumers choose certain products, channels and competitors over others (Faulds et al. 2018). The model states that, consumer will mostly run through seven stages before a consumption decision is made.

CDP modeling is a combination of traditional market research and unique quantitative modeling. It can help us to figure out the reasons behind the purchase. The seven steps are involved in CDP modeling, they are (a) need recognition; (b) search for information; (c) pre-purchase evaluation; (d) purchase; (e) consumption; (f) post-

---

[5]NIKE, Inc.—Inspiration and innovation for every athlete in the world; https://www.nike.com/hk.

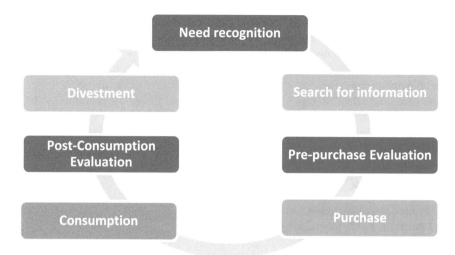

Fig. 15 Consumer Decision Process (CDP) model

consumption evaluation; and (g) divestment (Loudon and Della Bitta 1984; Hanna and Wozniak 2001; Harrell 1986) (Fig. 15).

Through research and investigation, company can understand more about how their targeted customers on their consumption decisions. These are very useful information to guide the company to develop an effective product and marketing mix.

(a) *need recognition*

The first stage is the need recognition. Consumption usually starts when needs are recognized. However, whether the consumption is realized, depends on how much discrepancy exists between the actual state (i.e. the consumer current situation) and the desire state (i.e. the situation in which consumer wants to stay). When this discrepancy exceeds a certain level or threshold, a need is recognized. However, if the discrepancy between the two states is below the threshold level, then need recognition will not take place (Blackwell et al. 1990).

Even though needs have been recognized, consumption may still not occur. The realizations of consumption rely on many factors. Firstly, we need to concern about the relative importance of the needs. For example, athletes may not feel the needs of a new compression sportswear since the old one still works, then they may not take the consumption action. In some case, although the athletes want to buy a new compression body suit, they may finally buy a pair of training tights instead as the price is comparatively lower.

To capture and increase market share, retailers need make their target consumers to recognize the needs of the product. In the case of sustainable compression sportswear, retailers also need to educate consumers upon the life spend and functions of the new products, especially when there is any new functional update, so to attract the consumer to purchase further even the old one still in a good condition.

Moreover, as products with sustainable features usually cost higher than the common ones, which could restrain consumers' purchase behavior. However, the need recognition of sustainable products could be inspired through promoting the planet wellbeing and sustainable lifestyles instead of individual functional features or physical products themselves. For example, sportswear brand Adidas, cooperate with an environmental initiative Parley for used recycle plastic from the ocean to produce a batch of products, including compression yoga tops, tights and trainers, etc. To promote this idea, they design a video with a slogan "Garbage in does not have to be garbage out". This brings the public awareness in marine life, and thus the need recognition of the selecting products is generated from them.[6]

(b) *search for information*

After need is recognized, consumer will try to gather information to search for potential satisfiers. In the second stage of the consumer decision process, it can be defined as motivated activation of knowledge stored in long term memory or acquisition of information from the environment. Searching can be either internal or external. Usually, people will search information from their memory. If the knowledge from the memory is good enough to provide a satisfactory course of action, then they will stop looking around for further information. However, this heavily depends on the adequacy or quality of their existing knowledge (e.g. their past experience and memory) (Van Staden 2011).

When internal search is inadequate, the consumer will turn to external search. External search can be defined as collecting additional information from the environment or marketplace. The motivation behind the pre-purchase search is the desire to make better consumption choice. Especially in the situation of buying sustainable compression sportswear, the users may not have enough knowledge upon the key feathers i.e. compression sportswear functions and the level of sustainability. They may question, how can compression sportswear help athletes to achieve better performance? If they want to recover faster, how long they need to wear after exercises? How to certify the level of sustainability of the product? In order to purchase the most suitable items, external information search is terribly needed and thus the degree of search is definitely high (Blackwell et al. 1990).

For sportswear brands like Asics, athletes can find the muscle support guide from their official website to have a better understanding upon the mechanism of the compression sportswear on how it works and what items they provide corresponding to the function descriptions.[7] Meanwhile, Asics provide vast information upon their commitment on sustainability. Not limited to the sustainable materials they used for the products, they consider the suppler chain management and the pollution during the production. All this kind of information can help consumer understand upon the product and the brand.[8]

---

[6]Adidas United States; http://www.adidas.com/us/parley.

[7]Muscle Support Guide: Run Further with Compression | ASICS; http://www.asics.com/gb/en-gb/running-advice/muscle-support-explained/.

[8]ASICS Corporate Social Responsibility and Sustainability; https://corp.asics.com/en/csr.

(c) *pre-purchase evaluation*

During this pre-purchase evaluation stage, consumer must determine the evaluation criteria used for judging alternatives, and decide which alternatives to be assessed on the performance of the considered alternatives, finally select and apply a decision rule to make the final choice (Robinson and Doss 2011).

When the consumers gather sufficient information, they may set up some criteria for selecting the alternatives. For example, the brand reputation, price of the production line, quality of the product. Referring to the criteria, they can figure out some alternatives which may satisfy their needs. By selecting the method of selection (for example to kick out the choices which cannot meet any one of the criteria or select one which can fulfill all the requirements), consumers can make their final consumption decision. The likelihood of a product being purchased depends on whether it is evaluated favorably by consumers.

(d) *purchase*

Purchase is the fourth stage of the consumption decision process which is the action stage. It is a very important as it is the only stage which generates actual sales. Purchase are often fully planned by going through all the previous stages, the consumers gather and process the information; and it is time for them to take action of purchasing the selected goods. Making use of the consumer behavior, purchase is a function of two factors: (1) purchase intention, and (2) environmental influence and/or individual difference. The level of influence and intention varies in different buying situations. Most of the purchase is fully planned but, in some cases, purchase purely bases on impulse which can be spontaneous and hedonic motivation.

Besides, a purchase can be made either at retail store or via on-line platforms. If the purchase occurs in the sportswear retail store, it provides a great chance for retail brands to directly contact the consumers in the sales floor and to build up positive relationship by providing face to face service. With technical and logistics advancement, on-line sales are also popular. Consumers can buy the product through internet and the goods will be delivered directly to their home or any desired places. Some brands, e.g. Puma, provide a 30-day return policy if consumers buy from Puma.com. This policy can facilitate consumer purchase as they do not need to worry if there exist fitting problem upon the sportswear they purchased.

(e) *consumption*

Consumption is the fifth stage; it starts right after the consumer using the product. Once the consumer uses the good, the utility of the product starts depreciating. During this consumption process, the consumer can experience the usefulness of the product and then outcomes will be used to compare with the expectation in the following stage. Consumption stage is critically important in the consumer decision making process, as it will affect the consumers' attitude towards the retail brand. If the consumption experience is not satisfied, re-purchase will not occur. This will seriously harm the brand's long-term business. Under Armour is an upcoming sportswear brand which aims at providing high performing, high quality, extremely durable and comfortable products to their customers. Their products strive to help sports players to perform better, and also ensure consumers to use longer. A satisfactory consumption experience helps the brand build up positive image among the sports players, and also develop a long term relationship with them, facilitating repeat purchase in the future.

(f) *post-consumption evaluation*

Evaluation upon the consumed products or goods is necessary, which can help the consumer know if unwise choice had been made or can avoid in the future. Every consumer has a certain expectation about the product or service to be provided before purchase. There are different types of expectations, including (1) equitable performance: a normative judgment reflecting the performance that one ought to receive given the cost and efforts devoted to purchase and use; (2) ideal performance: the optimum or hope- for "ideal" performance level; and (3) expected performance: what the performance probably will be. The post-consumption evaluation is to evaluate the difference between the expected performance of the product and the actual outcomes after consumption. The judgment takes three different forms: (1) positive disconfirmation: performance is better than expected; (2) simple conformation: performance equals expectation; and (3) negative conformation: performance is worse than expectation. This experience and evaluation outcome will be stored in their memory, which will be internal information for future consumption. This information will greatly affect the intention of the consumers on their repeat buying.

(g) *divestment*

When time goes by, the utility of the product being consumed will decrease. Divestment will occur when the product depreciates to a point that the utility is 0. Consumers have serval options, which include the outright disposal, recycling or re-marketing. If the consumer still need the same kind of product, he/she will concern if buying the same product or switch to another alternative. The growth of fast fashion increases the clothing disposal, which bring high pressure to the environment and also bring up high awareness of sustainable fashion. Sportswear brand Patagonia encourage their customers to send back the unused products and promote sustainability concept. The brand will repair the products and thus old Patagonia clothing will not end up in a landfill or an incinerator. When the consumers send the old cloths to their stores, the brand could get chance to understand why consumer dispose the product (e.g. any dissatisfaction or seek for new functionality, etc.), thus they can provide consumers better products and services to encourage their re-peat buying in future.

5 Concluding Remarks

Due to the growth of healthy lifestyle and environmental concern, sustainable sportswear has become a promising fashion trend. In this chapter, we introduced the present status and future development of sustainable sports fashion, and explore how compression sportswear can fit into this sustainable sportswear trend by integrating the 5Ps model and triple bottom line. Compared with traditional fashion apparel (woven, sweater, cut and sew knit), the product design and production process of seamless sportswear would be an interesting area in sustainable development. Integrating with the computer aided design and manufacturing (CAD/CAM), seamless knitting manufacturing process largely reduce fabrication steps and material wastes in prototyping and mass production process, which not only provides consumers desirable functional properties, but also carries sustainable features to cater to the sustainable sportswear trend. Followed that, we further discussed the corresponding features and requirements that consumers will take into consideration in consumption of functional sportswear, including consumers' attitudes and influencing factors. At the end, we further explored how the sportswear brands can increase sales by applying consumer decision process (CDP) model to promote repeat-purchase relating to sustainable sports fashion.

Acknowledgements The authors would like to thank the Hong Kong Polytechnic University to support this study through postgraduate programme 88011 and research project 1-ZVLQ.

References

Arnould, E. J., Price, L., & Zinkhan, G. M. (2004). *Consumers* (2nd ed.). New York, NY: McGraw-Hill/Irwin.

Berry, M. J., & Mcmurray, R. G. (1987, June). Effects of graduated compression stockings on blood lactate following an exhaustive bout of exercise. *American Journal of Physical Medicine, 66*(3), 121–132.

Blackwell, R. D., Miniard, P. W., & Engel, J. F. (1990). *Consumer behaviour* (6th ed.). Harcourt College.

Burger, J. (1993). *Personality* (3rd ed.). Pacific Grove, Calif.: Brooks/Cole Pub.

Catalyst Corporate Finance. (2014, Spring). Global Sportswear Sector M&A update.

Cruz, J. M., & Wakolbinger, T. (2008). Multiperiod effects of corporate social responsibility on supply chain networks, transaction costs, emissions, and risk. *International Journal of Production Economics, 116*(1), 61–74.

Davies, V., Thompson, K. G., & Cooper, S.-M. (2009). The effects of compression garments on recovery. *Journal of Strength and Conditioning Research, 23*(6), 1786–1794.

de Brito, M. P., Carbone, V., & Blanquart, C. M. (2008). Towards a sustainable fashion retail supply chain in Europe: Organisation and performance. *International Journal of Production Economics, 114*(2), 534–553.

Duffield, R., Cannon, J., & King, M. (2010). The effects of compression garments on recovery of muscle performance following high-intensity sprint and plyometric exercise. *Journal of Science and Medicine in Sport, 13*(1), 136–140.

Engel, F., & Sperlich, B. (2016). *Compression garments in sports: Athletic performance and recovery*. Cham: Springer International Publishing.

Faulds, D. J., Mangold, W. G., Raju, P. S., & Valsalan, S. (2018, March–April). The mobile shopping revolution: Redefining the consumer decision process. *Business Horizons, 61*(2), 323–338.

Fishbein, M., & Ajzen, I. (1975). *Belief, attitude, intention, and behavior: An introduction to theory and research*. Reading, MA: Addison-Wesley Pub. Co.

Gallopín, G. C., & Raskin, P. D. (2002). *Global sustainability bending the curve*. London: Routledge.

Hanna, N., & Wozniak, R. (2001). *Consumer behavior: An applied approach*. Upper Saddle River, N.J.: Prentice Hall.

Hannouf, M., & Assefa, G. (2017). Life cycle sustainability assessment for sustainability improvements: A case study of high-density polyethylene production in Alberta, Canada. *Sustainability, 9*(12), 2332.

Harrell, G. D. (1986). *Consumer behavior*. San Diego: Harcourt Brace Jovanovich.

Hayes, S. G., & Venkatraman, P. (2016). *Materials and technology for sportswear and performance apparel*. Boca Raton, FL: CRC Press.

Hiller Connell, K., & Kozar, J. (2017). Introduction to special issue on sustainability and the triple bottom line within the global clothing and textiles industry. *Fashion and Textiles, 4*(1), 1–3.

Jung, Y. J., & Kim, J. (2016, July 2). Facebook marketing for fashion apparel brands: Effect of other consumers' postings and type of brand comment on brand trust and purchase intention. *Journal of Global Fashion Marketing, 7*(3), 196–210.

Karaosman, H., Morales-Alonso, G., & Brun, A. (2016). From a systematic literature review to a classification framework: Sustainability integration in fashion operations. *Sustainability, 9*(1), 30.

Katz, D. (1960, July 1). The functional approach to the study of attitudes. *The Public Opinion Quarterly, 24*(2), 163–204.

Kraemer, W. J., Bush, J. A., Newton, R. U., Duncan, N. D., Volek, J. S., Denegar, C. R., et al. (1998, February 1). Influence of a compression garment on repetitive power output production before and after different types of muscle fatigue; Taylor & Francis Group; *Sports Medicine, Training and Rehabilitation, 8*(2), 163–184.

Lee, T. S. (2003). *A study of consumer's self and purchasing behaviour in fashion brand image marketing* (Ph.D. Thesis). The Hong Kong Polytechnic University.

Li, M., Wang, X., Wang, T.-T., & Yang, Y.-X. (2010). The influencing factors of consumer's purchase decision-making of fast fashion brands. *Donghua Daxue Xuebao (Ziran Ban)/Journal of Donghua Univeristy, 36*(2), 208–212, 228.

Liu, R., & Little, T. (2009). The 5Ps model to optimize compression athletic wear comfort in sports. *Journal of Fiber Bioengineering and Informatics, 2*(1), 41-51.

Liu, R., Little, T., & Eugene W. M. (2012). Evaluation of elite athletes psycho-physiological responses to compression form-fitted athletic wear in intensive exercise based on 5Ps model. *Fibers and Polymers, 13*(3), 380–389.

Loudon, D. L., & Della Bitta, A. J. (1984). *Consumer behavior: Concepts and applications* (2nd ed.). New York: McGraw-Hill.

Lovell, D. I., Mason, D. G., Delphinus, E. M., & Mclellan, C. P. (2011, December). Do compression garments enhance the active recovery process after high-intensity running? (Author abstract) (Report). *Journal of Strength and Conditioning Research, 25*(12), 3264(5).

Lynn, R. K. (1984). *Attitudes and social adaptation: A person-situation interaction approach*. Oxford: Pergamon Press.

Marqués-Jiménez, D., Calleja-González, J., Arratibel-Imaz, I., Delextrat, A., Uriarte, F., & Terrados, N. (2017, October 29). Influence of different types of compression garments on exercise-induced muscle damage markers after a soccer match. *Research in Sports Medicine*, 1–16.

Martínez, S., Errasti, A., & Rudberg, M. (2015, February 19). Adapting Zara's 'Pronto Moda' to a value brand retailer. *Production Planning & Control*, 1–15.

Molkenthin, M. E. (2012). *"It's so me!" using value-expressive and socially-adjusted attitude functions to predict counterfeit purchase intention* (M.S. thesis). University of South Carolina.

Muzikante, I., & Reņģe, V. (2011). Attitude function as a moderator in values-attitudes-behavior relations. *Procedia - Social and Behavioral Sciences, 30*, 1003–1008.

Park, H., & Kim, Y.-K. (2016). An empirical test of the triple bottom line of customer-centric sustainability: The case of fast fashion. *Fashion and Textiles, 3*(1), 1–18.

Phares, E., & Chaplin, W. (1997). *Introduction to personality* (4th ed.). New York: Longman.

Piras, A., & Gatta, G. (2017, June). Evaluation of the effectiveness of compression garments on autonomic nervous system recovery after exercise (Report) (Author abstract). *Journal of Strength and Conditioning Research, 31*(6), 1636(8).

Rannikko, A., Harinen, P., Torvinen, P., & Liikanen, V. (2016, September 13). The social bordering of lifestyle sports: Inclusive principles, exclusive reality. Routledge; *Journal of Youth Studies, 19*(8), 1093–1109.

Reynolds, G. (2016, July 29). Ask well (Science Desk) (WELL). *The New York Times*, D4(L).

Robinson, T., & Doss, F. (2011, July 12). Pre-purchase alternative evaluation: Prestige and imitation fashion products. *Journal of Fashion Marketing and Management: An International Journal, 15*(3), 278–290.

Rowley, J. (1997, March 1). Knowing your customers. *Aslib Proceedings, 49*(3), 64–66.

Samli, A. C. (2013). International consumer behavior in the 21st century impact on marketing strategy development. New York, NY: Springer.

Shen, B. (2014). Sustainable fashion supply chain: Lessons from H&M. *Sustainability, 6*(9), 6236–6249.

Shen, B., Li, Q., Dong, C., & Perry, P. (2017). Sustainability issues in textile and apparel supply chains. *Sustainability, 9*(9), 1592.

Szmydke, P., & Marfil, L. (2015, April 10). H&M: Sustainability a force behind long-term success. *WWD, 209*(73).

Van Staden, J. (2011). Information seeking by female apparel consumers in South Africa during the fashion decision-making process. *International Journal of Consumer Studies, 35*(1), 35–49.

WCED (World Commission on Environment and Development). (1987). *Our common future*. Oxford: Oxford University Press; New York: UNCED (United Nations Conference on Environment and Development), 1992: Agenda 21, UN.

Yeager, S. (2010). 2010 is… The big squeeze. *Bicycling, 51*(3), 24.

Yolanda, A., Nurismilida, & Herwinda, V. (2017, June 1). Affect of cultural factor on consumer behaviour in online shop. *International Journal of Scientific & Technology Research, 6*(6), 287–292.

Yurchisin, J., & Johnson, K. (2010). *Fashion and the consumer* (English ed., Understanding fashion series). Oxford; New York: Berg.

The Impact of Knowledge on Consumer Behaviour Towards Sustainable Apparel Consumption

Nazan Okur and Canan Saricam

Abstract The concept of sustainability in apparel involves designing, manufacturing and consuming of products with the consideration of environmental and social impacts specifically. Today, apparel retailers do not only compete to offer a variety of goods to the consumers, but also compete to implement sustainability into their business models. From the consumers' perspective, the ever increasing launch of environmentally friendly products/brands manufactured with an understanding of corporate social responsibility is encouraging them to take a more active role in sustainability via their apparel purchasing. As a result of the increased awareness, consumers are more likely to purchase sustainable products; however, knowledge is the key issue triggering the consumers towards sustainability in apparel purchasing. That is to say; if consumers know more about sustainability, then they are more likely to purchase sustainable apparel products. This chapter deals with the empirically testing of a model developed for the investigation of consumers' behaviour towards sustainable apparel consumption including the constructs of knowledge of environmental issues, knowledge of social issues, motivation for environmental responsibility, attitude toward green brand and purchasing intention. A survey was conducted among 796 participants in Turkey. Confirmatory factor analysis (CFA) was conducted to test the validity of the items of each construct, and structural equation modelling (SEM) was used to estimate the relationships between the constructs. The results revealed that motivation for environmental responsibility is strongly correlated with the consumers' attitude toward green brand. Moreover, knowledge of consumers about environmental issues was found to have a significant impact on purchasing intention. On the other hand, knowledge of social issues was found to have an insignificant effect on consumers' sustainable apparel purchasing intention.

N. Okur (✉) · C. Saricam
Department of Textile Engineering, Faculty of Textile Technologies and Design, Istanbul
Technical University, Inonu Cad. No: 65 34437, Beyoglu, Istanbul, Turkey
e-mail: okurn@itu.edu.tr

C. Saricam
e-mail: saricamc@itu.edu.tr

© Springer Nature Singapore Pte Ltd. 2019   69
S. S. Muthu (ed.), *Consumer Behaviour and Sustainable Fashion Consumption*,
Textile Science and Clothing Technology,
https://doi.org/10.1007/978-981-13-1265-6_3

Keywords Sustainable · Apparel · Purchasing · Knowledge · Environment
Social

1 Introduction

The term sustainability can be simply defined as meeting the needs of the present
without compromising the needs of the future generations (Brundtland Report 1987).
Over the years, concept of sustainability has expanded and in today's conjuncture,
it is commonly regarded as the basis of three issues as Society, Environment and
Economy; or in other words, it is regarded as a way of creating value for People, Planet
and Profit (Thiele 2014). The apparel industry has also been greatly influenced by
the expansion of sustainability concept and the dynamics of the industry has changed
both for the business side and consumer side. Cleaner production with less impact on
the environment and producing less waste can create a value for Planet, safe and good
working conditions can create a value for People and emerging new business models
that represent the vision of sustainability can create Profit (Niinimäki 2013). Thus;
the concept of sustainability in apparel involves designing, manufacturing, logistics,
retailing, consuming and disposal of products with the consideration of specifically
environmental, social and economic impacts.

There are several factors and motivation behind consumer's attitude and
behavioural intention towards sustainable apparel consumption. Among them,
knowledge of consumers about sustainability concerns plays a key role and hence,
it has been a focus of several research.

This chapter deals with overviewing the sustainability concepts with regard to
designing, production and consumption of apparels as well as measures of apparel
sustainability. The chapter discusses knowledge of sustainability in terms of environmental and social issues and its influences on consumers' sustainable apparel
purchasing behaviour. In addition, this chapter proposes an empirically testing of
a model developed for the investigation of consumers' behaviour towards sustainable apparel consumption including the constructs of knowledge of environmental
issues, knowledge of social issues, motivation for environmental responsibility, attitude toward green brand and purchasing intention.

2 Sustainability of Fashion and Apparel Products

The term sustainability was first basically defined in Brundtland Report by The World
Commission on Environment and Development in 1987 as meeting the needs of the
present without compromising the needs of the future generations (Brundtland Report
1987). Afterwards, the definition of sustainability has been built on triple bottom line
concept, which includes environment, society, and the economy (Fig. 1).

Fig. 1 Three nested
dependencies model in
sustainability (Willard 2012)

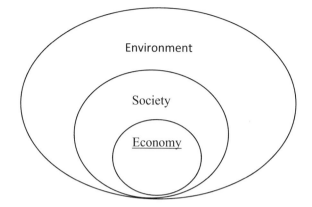

Environmental Sustainability: Sustainability is usually considered in terms of environment by most of the people. The reason for this intention might be originated due to the first usage of the word "Sustainability" in an environmentalist text "A Blueprint for Survival" written by Edward Goldsmith in 1972 (The Ecologist 1972). Also according to Wiersum (1995) the approach of sustainability was fundamentally based on forestry that covers the idea of not to harvest more than the forest yields.

Today, environmental sustainability is considered as the most global issue than the other two branches of sustainability (economic and social) due to its issues are in direct connection with our planet and all livings on it. This connection can also reach out the future of posterity. The corresponding issues of ecological (i.e. environmental) sustainability covers temporary or permanent variations of lands, water resources and the atmosphere due to human activities, which may generate either reversible or irreversible impacts on our nature.

Environmental dimension of a sustainable apparel product includes the use of renewable resources, effective recycle of the materials, minimization of waste production and treatment of waste without a hazardous impact on the environment, use of energy-saving production principles and technologies (Niinimäki 2013).

Economic Sustainability: In today's world, the concept of economic sustainability is concerned with maintaining capital secured, making use of capital sources with the best possible optimization to provide beneficial balance over the long term. In this context, approach of economic sustainability may seem as the optimization of capital sources but additionally economic sustainability is also dealing with environmental and social aspects of sustainability or sustainable development. Economic and ecological sustainability has a common focus area that covers the effects of over manufacturing on natural sources and, use of capital sources for preserving natural sources and protect environment. In the intersection of economic and social sustainability or sustainable development, the main aim is to ensure an increase of general welfare of the people (Garbie 2016).

Social Sustainability: The concerns of social sustainability are based on labour and working conditions, occupational health and safety, land acquisition and resettlement, indigenous peoples and community health, safety and security. Currently, the role of social sustainability is abnegated in comparison with the other two branches. The most recent approach of social sustainability favours that all domains of sustainability or sustainable development are dependent on social sustainability. James (2014) emphasized this relation as it is ensured with all human activities which are in contact with environment and economy.

Social dimension of a sustainable apparel product involves primarily the labour conditions and socially responsible practices of the companies. Finally, financial feasibility is accordingly an important aspect of sustainable apparel product in terms of economic dimension (Niinimäki 2013).

The existing system in the apparel industry is based on the rapid cycle of fashion trends that aim at continuous production, new customer needs, and new products. The life cycle of products has shortened and retailers and brands want their products to be changed at a higher rate (Niinimäki and Hassi 2011). In addition, low prices have created more buying habits in the consumer market.

In the past periods, new products were offered to the consumer market for only two seasons a year and the main street stores made two products stock per year, including autumn/winter, spring/summer. However, in today's apparel consumer market, new products are offered almost every week. Essentially, this is the result of the "fast fashion" phenomena, which was emerged in 1980s and has spread all over the world since then. The leading brands that use fast fashion as their business model are Zara, Benetton, GAP and H&M. Fast fashion integrates the consumer preferences into design process in a very short time interval and offers trendy and fashionable products to the consumer market quickly (Gabrielli et al. 2013). The term "Fast" in this business model refers to the mass production of apparel products in a very short time, offering the products with affordable prices and shortening the life-span of a product. Now consumers are able to buy more apparel commodities with a smaller budget. On the other hand, this type of fashion concept has been questioned and required to be rethought due to some environmental and social concerns. Primarily, this type of production uses more material resources and labour force, and the use of more material resources results in increasing pollution on the environment. According to figures in 2015, 79 billion cubic meters of water was consumed in the fashion industry. This created 1715 million tonnes emissions of CO_2 and 92 million tonnes of waste. For the year 2030, water consumption of the fashion industry is projected as 118 billion cubic meters with 50% increase, whereas emissions of CO_2 and the amount of waste that will be created in 2030 are projected as 2791 million tonnes and 148 million tonnes respectively, with around 60% increase (Pulse of the Fashion Industry 2017). Moreover, since the concept of fast fashion has changed the perception of consuming and fashion from consumers' perspective, it has created an addiction for shopping and overconsumption. Thus, throwaway or disposable fashion (Bhardwaj and Fairhurst 2010) the amount of apparel disposal, which are rarely biodegradable, has increased. It is a significant threat to the environment, because the vast majority of apparel waste ends up in landfills or is destroyed; and globally, only 20% of apparel is collected for reuse or recycling (Pulse of the Fashion Industry 2017; The Ellen MacArthur Foundation 2014).

Enabling to offer products at affordable prices requires cost cutting along the supply chain. This may lead to depressed wages and unwarranted working conditions. This is the primary concern about the sustainability of fast fashion concept with regard to social issues.

The concerns about the sustainability of existing business models, particularly fast fashion model, in the apparel industry has forced the actors of the industry to consider sustainability from the fibre used in the production to the store, and event post-consumption of the apparel products and forced them to transit sustainable business model and develop sustainable business strategies.

In the 1990s, sustainability was emphasized in trend forecasts of natural and recycled fibres, industry-related publications and fair shows. Since 2000, however organic products, fair trade and renewable fibres have basis of innovations and many companies have created their collections using alternative materials. In recent years, there has been a great interest that has begun to emerge in ecological matter and textiles in Paris, London and New York. Now, designers, producers and consumers seem to want to get information about creative solutions and precautions that can be taken about the evolution of textiles and their impact on the environment (Türkmen 2009; Can and Ayvaz 2017). Thus, the primary practices have involved the use of natural and biodegradable materials instead of synthetics and replacement of hazardous chemicals with environmentally friendly constituents (Jung and Jin 2014; Can and Ayvaz 2017). Further, the reduction of use the of resources and reduction of waste through the recycling have been incorporated to the sustainability approach in fashion and apparel industry (Jung and Jin 2014; Can and Ayvaz 2017). In other words, instead of a linear supply chain model, a circular business model has gained importance (The Ellen MacArthur Foundation 2014).

Sustainability in fashion and apparel industry covers such topics that are abandoning the use of artificial and synthetic fibres completely or decreasing very low levels, using recyclable and naturally soluble materials, increasing the product lifespan by means of strength and classical design, introducing new design and production processes, designing less and more intelligent clothes, promoting sustainable agriculture and addressing the ethics of the fashion industry (Gürcüm and Yüksel 2012; Can and Ayvaz 2017; Black 2008).

The aim in the sustainable fashion is to create systems that can be maintained indefinitely and to play along with environmentalism and social responsibility principles. Sustainable fashion can also be positioned as a subset of sustainable design concept. The advantages of sustainable fashion are as follows (Mangır 2016; Can and Ayvaz 2017):

- Improving environmental and social performance will increase the popularity of the brand
- Participation of companies to the social and environmental projects will lead to stronger ties with consumers
- The use of technology in the production chain will reduce the costs of the companies

- Presence of many environmentally conscious brands will increase the environmental performances of companies.

With regard to sustainable apparel product, it covers the environmental, social/ethical and economic dimensions of the product in production, use and post-consumption stages (Tischner and Charter 2001). Within a wider concept, it covers the designing, production, logistics, retailing, consuming and disposal of a product.

Although the terms fashion and sustainability seems to be as contradictory phenomena, sustainability can be certainly incorporated into fashion business models. For instance, being not actually the opposite of fast fashion, recently slow fashion concept has been emerged an alternative business model that represents a vision of the sustainability in the industry of the fashion (Capdevila 2014). Following the slow food and slow cities approaches inspired within the perspective of slow movement, the world of the fashion began worrying for the philosophy Slow and in 2007 the researcher Kate Fletcher defined the "Slow fashion" as being about designing, producing, consuming and living better. Slow fashion is not time based but quality based (Fletcher 2007).

Clark (2008) defines three goals of slow fashion as valuing local resources, transparent production systems and creating sustainable products. The slow fashion concept rejects the large-scale, speed up models that are the standard for the industry today. In the same way, slow fashion urges consumers to reconnect with an old way of living, encouraging them to take a more active role in their purchasing decision.

In the apparel collections that are designed with slow fashion approach; the products produced in the simplest forms with the finest quality fabrics are the examples of the use of the traditional weaving, embroidery or handcraft knowledge and skills. Slow Fashion means better designing, producing, consuming and a better life (Türkmen 2009; Can and Ayvaz 2017). Besides, the slow fashion movement aims to provide consumers to think about the origins and the materials of the clothes that they wear. Slow fashion creates a social responsibility by questioning what we buy, who produces the product, and how this information affects the quality of the product (Alpat 2013; Can and Ayvaz 2017).

The impact of designers and companies is crucial for ensuring sustainability in the fashion industry. The number of the fashion designers and companies that make sustainable based ecological design is gradually increasing. In 2013, 16 global brand and retailers—Nike, Adidas, Puma, H&M, M&S, C&A, Li-Ning, Zara, Mango, Esprit, Levi's, Uniqlo, Benetton, Victoria's Secret, G-Star Raw and Valentino—signed up to the Greenpeace Detox campaign, pledging to eliminate all hazardous chemicals throughout their global supply chains by 2020 (Stevenson 2013).

Although the biggest obstacle seems to be realized a sustainable fashion approach is the "fast fashion" approach; it is noteworthy that 'fast fashion' does not automatically represent a threat to the environment and the world economy according to the Pulse Fashion Industry Report that was produced by The Boston Consulting Group (BCG) and Global Fashion Agenda (GFA) in 2017 (Pulse of the Fashion Industry 2017). The report showed that the large-scale high-street retailers have higher Pulse scores (a score for measuring sustainability, which will be mentioned under the following heading "MEASURES OF APPAREL SUSTAINABILITY") than most of the market. Many large-scale entry-price high-street and sportswear retailers get strong Pulse Scores, as do the small "sustainability champions". On the other hand, most of the small and medium scale premium brands and retailers achieved moderate scores. Thus, it is concluded that the driver of the sustainability is the scale of the company rather that the price positioning.

Hence, although fast fashion, even only fashion, and sustainability are seemed to be contradictory phenomena at first glance; leading fast fashion retailers such as H&M can also be one of the leading brands that make sustainability one of the major components in their organization. H&M declares that sustainability in their organization is integrated into all parts of the organization—from every single department to all brands and markets. Besides, the company takes a long term view with regard to sustainability and avoids from short term solutions that do not lead to lasting change. Finally, the company believes to the importance of the universal and industry wide collaborations for sustainability. It collaborates with brands, trade unions, NGOs, experts and scientists, business partners and civil society will continue in terms of recycling innovations, new sustainable materials or the work to improve working conditions of the labourers (http://sustainability.hm.com/en/sustainability/about/about/ceo-message.html).

3 Measures of Apparel Sustainability

In order to validate the sustainability of an apparel product, participants in the industry have proposed some measuring tools and indices for evaluating the apparel products in terms of sustainability. The first step towards such a measuring tool was Nike's Environmental Apparel Design Tool (Nike 2010), which was followed by Outdoor Industry Association's OIA Social Responsibility Toolkit (OIA 2014). Nike's environmental design tool aimed to help designers in choosing the most suitable materials at the beginning of the design and production process. In this way, the tool is able to rate how apparel designs score in decreasing waste and increasing the use of environmentally friendly resources and materials while allowing the designers to make real time adjustments. Outdoor Industry Association's OIA Social Responsibility Toolkit is aiming to provide both retailers and suppliers with the tool to start and improve their corporate social responsibility program.

These attempts were then generalized to Higg Index that was developed by Sustainable Apparel Coalition. The Sustainable Apparel Coalition is the apparel, footwear, and textile industry's leading alliance for sustainable production. Beyond being just a measuring tool for sustainability, Higg Index helps the industry participants to determine inefficiencies, resolve damaging practices, and work to achieve the environmental and social transparency that consumers are demanding (https://apparelcoalition.org/the-sac/).

In its current practise, Higg Index includes product, facility and brand and retail tools that measure environmental and social labour impacts across the supply chain.

Higg product tools can be used during a product's design phase to understand its predicted impact. They offer brands and manufacturers information to make better choices at every stage of a product's development (https://apparelcoalition.org/higg-product-tools/). With regard to Higg facility tools, it can be identified as the tools to measure the social and environmental performance of the facilities. Benchmarking of the performance of a facility against that of their peers is also possible by using Higg facility tools (https://apparelcoalition.org/higg-facility-tools/). In addition, Higg brand and retail tools enable the brands and retailers to measure the environmental and social impacts of their operations and make significant improvements. These tools also allow brands and retailers to share sustainability information with key stakeholders, including supply chain partners. The environmental impacts that are measured include greenhouse gas emissions, energy use, water use and water pollution, deforestation, hazardous chemicals and animal welfare; whereas social impacts that are measured by using Higg brand and retail tools include use of child labour, discrimination, use of forced labour, sexual harassment and gender based violence in the work place, non-compliance with minimum wage laws, bribery and corruption, working time, occupational health and safety and responsible sourcing (https://apparelcoalition.org/higg-brand-tool/).

Based on the Higg Index and aimed at extending its scope to extrapolate its findings to the whole fashion industry; Pulse score was developed as a performance score for measuring and tracking the sustainability of the global fashion industry on key environmental and social impact areas measured on a scale from 1 to 100 (Pulse of the Fashion Industry 2017). The Pulse Score is mainly based on Higg Index's brand tool and it is complemented by expert interviews that are conducted with sustainability managers, a survey answers to reconfirm sustainability patterns and performance to increase sample size and fair market representation further, and expert sounding board to validate and discuss the results (Pulse of the Fashion Industry 2017). Including the stages of design and development, raw materials, processing, manufacturing, transportation, retail, use and end of use of the apparel product, overall Pulse Score of the fashion industry is 32. When these stages of the value chain are evaluated on their own, it is seen that processing and transportation stages get the highest score of 38 and 41 respectively; whereas end-use of the apparel products and raw materials stages get the lowest score of 9 and 17 respectively. Design and development, use of apparel products, manufacturing and retailing stages get moderate scores, as 22, 23, 28 and 28 respectively (Pulse of the Fashion Industry 2017).

When the Pulse Scores of the brands are evaluated with regard to their geographic origin, it is indicated that European brands score better along environmental issues, whereas US brands are more compliant on social and labour practices (Pulse of the Fashion Industry 2017).

For measuring the consumers' knowledge about sustainability and their attitude and behaviour towards sustainable apparel consumption methodically, several scales have been developed by researchers to date.

Assessing consumers' knowledge of environmental issues in the manufacture and recycle of apparel products, Kim and Damhorst (1998) developed the "Environmental Apparel Knowledge Scale" that consist of 11 items. The scale basically points at use of natural fibres for protecting environment, hazardous influences of production processes of manufactured fibres, textile finishing processes and compliance of textile companies to governmental legislations for clean air and water. Later, Dickson (1999) developed "Knowledge of and Concern with Apparel Social Issues Scale" to determine consumers' knowledge about social issues executed in clothing manufacturers operating both in US and foreign countries. The practices of the clothing manufacturers with regard to social issues include the use of child labour, payment of local minimum wage, reasonable working hours, providing safe workplaces for the employees. The scale also includes items for measuring consumers' attitudes toward apparel social issues. In 2000, Dunlap et al. (2000) developed "New Environmental Paradigm Scale" for measuring consumers' general attitudes toward human and environment interactions. The 15 items included in that scale are designed to evaluate the people's opinion about people's impact on environment, their right to modify the natural environment, the strength and the capabilities of the nature and the balance of the nature.

Kozar and Hiller Connell (2010) developed "Sustainable Apparel Purchasing Behaviour Scale" for measuring environmentally and socially sustainable apparel purchasing behaviours. The items measuring environmentally sustainable apparel purchasing behaviours point at the apparel purchases of the consumers in the past considering the environmental policies or practices of the apparel brand or retailers. The items measuring the socially sustainable apparel purchasing behaviours point at the apparel purchases of the consumers in the past considering the manufacturing of the apparel products whether they are produced in a sweatshop or because the workers were treated unfairly.

4 Knowledge About Sustainability in the Apparel Industry

4.1 Knowledge About Environmental Issues

Up to date, a number of studies have been indicated that the level of consumers' knowledge about sustainability is directly related to the consumer environmentally friendly behaviour and consequently influences willingness to purchase sustainable apparel products.

D'Souza et al. (2006) stated that if a consumer has knowledge about the environment and pollution pronouncement, the causes and impact on the environment, then their consciousness would increase that would stimulate an encouraging attitude towards green products.

According to Lin (2009), the consumers who had organic cotton products in the past are more likely to purchase organic cotton products and they are and more willing to pay higher prices for other organic products. Besides, a positive correlation was found between consumers' knowledge about sustainability and purchasing apparel made of organic cotton (Kang et al. 2013). Therefore, it can be stated that when the consumers have more information about environmentally sustainable apparel products and have experienced such kind of products, they can develop positive approach and attitudes towards consumption of environmentally sustainable apparel products, which trigger them to purchase that kind of products (Kang et al. 2013). Hustvedt (2006) found out that the organic cotton consumers are concerned about the environmental effect of apparel manufacturing and are certain that organic agriculture is good for the environment. Moreover, Hustvedt and Bernard (2008), searched for consumers' willingness to pay for value based labelling of the products for three attributes of fibre as origin, type and production method. For this aim, they conducted an experimental survey among the students for socks with varying information. They found that the payments for organic and non-genetically modified fibres are similar with labelling of organic cotton being cost slightly more to consumers. Therefore; they concluded that additional marketing activities are not required about the non-genetically modified features of organic fibres. On the other hand, the results of the study revealed that consumers value information about the local origin of fibres. As a result, it can be suggested that sustainable production systems that are not organic may be successful if they emphasize other features such as local origin or non-genetically modified production.

Saricam and Erdumlu (2010) investigated the influence of knowledge, environmental sensibility and budget of a group of university students in Turkey on purchasing motivation and purchasing decisions for environmentally friendly apparel. The knowledge related items in the survey conducted were including materials used in the production of environmentally friendly apparel, processes, chemicals used and the waste occur during the production. Half of the participants were attending to the textile engineering program, while the other half the participants were attending to the mechanical engineering program. The results of the study revealed that there is a significant difference between the knowledge of students about environmentally friendly apparel products. The textile engineering students were found to be more knowledgeable about sustainable apparel based on their background in their educational area. The difference between the answers of two group of students were analysed statistically and found to be significant.

While attempting to investigate the purchasing attitude and intention of consumers towards sustainable apparel products; most of the time, knowledge of the consumers about environmental and social issues related to sustainability in apparel consumption is disregarded, and the consumers in consideration are recognized as already concerned and have knowledge about sustainability in apparel consumption. On the other hand, substantially, it is important to determine how the consumers define sustainability in general and in the apparel industry, what they know about the impact of apparel consumption on environment and also social and economic issues; for establishing more efficient marketing strategies.

The level of knowledge about sustainability in the apparel industry and its impacts on purchasing intention can differ for consumers in different generations. So far as the knowledge of sustainability held by young consumers is explored, it was realized that the students relate the sustainability initially to environmental issues and then long-term effects; next to social and finally to economic issues (Carew and Mitchell 2002; Gam and Banning 2011; Kagawa 2007).

4.2 Knowledge About Social Issues

While environmental concerns about sustainable clothing and apparel products, such as water and energy consumption, use of hazardous materials (pesticides and chemicals) and high amount of waste production, were more commonly being discussed among consumers in the early stages of sustainable apparel design and production; later sustainability concept has also been started to be linked to the fair traded products. In addition to the socially responsible practices and implementations of the companies; consumers have started to think about the role of recycling and vintage in sustainable clothing as the knowledge and experience of the consumers about sustainability have evolved (Cervellon and Wernerfelt 2012).

According to Stephens (1985), consumers who had knowledge regarding the corporate social responsibility issues in the apparel industry were more likely to involve in socially responsible apparel purchasing behaviour.

The knowledge of consumers about the applications of the apparel companies for social responsibility has been examined so far, and it has been realized that there is a high level of consumer mindfulness about companies' social responsibility practices, particularly in terms of labour concerns such as the use of child labour and not paying fairly wages (Dickson 1999; Kozar and Hiller Connell 2010). Kozar and Hiller Connell (2010) stated that more than half of the consumer included in their research was knowledgeable about socially responsible practices of apparel companies. Besides, almost two-thirds of the consumers felt that their awareness regarding domestic and foreign apparel manufacturing was high. In a survey conducted among US consumers to assess the relationship between consumers' attitudes about social issues and their socially responsible purchasing behaviours, it was found out that there is a positive relationship consumers' attitudes toward people living in developing nations and purchase intentions towards socially responsible apparel companies (Kim et al. 1999).

With regard to ethical issues that companies have to deal with or challenge, it has been obvious for a period of time that the consumers can easily get information about fair or unfair practices of the companies via several sources such as their own experiences, mass media, social media, and word of mouth and this information builds the consumers' expectations about apparel companies' ethical behaviour (Thompson 1995; Creyer and Ross 1997; Dickson 2000; Black 2008; Dickson 2001; Hustvedt and Dickson 2009).

Nevertheless, when the attitude or purchasing intention of the consumers towards a socially responsible apparel brands is evaluated, it was indicated by Carrigan and Attalla (2001) and Page and Fearn (2005) that price, quality and value of the brand are considered more important by the consumers rather than practices of the brand in terms of social responsibility. In fact, consumers sometimes may perceive that sustainable apparels are more expensive or they are out of style in spite of their related knowledge about sustainability and attitudes towards sustainability (Kim and Damhorst 1998).

4.3 Knowledge of Green Brands

According to Keller (1993), green brand knowledge can be related to the information about the environmental commitment and environmental concerns about a brand. Consumers expect to get consistent information about environmental concerns to enhance their green brand knowledge and to ease their green product purchases. Within the literature, there are several findings revealing the positive relationship between knowledge of consumers on environmental concerns and consumers' purchase intention of green products (Pagiaslis and Krontalis 2014; Peattie 2010; Yadav and Pathak 2016; Chen and Chang 2012; Eze and Ndubisi 2013). Observing the green consumer behaviour in Asian markets, Kumar and Ghodeswar (2015) found that supporting environmental protection, drive for environmental responsibility, green product experience, environmental friendliness of companies and social appeal are identified as significant variables influencing green product purchase decisions of the Indian consumers. On the other hand, these findings were related to the knowledge of green brand in general, not restricted to the apparel brands. Kang et al. (2013) conducted a survey among students attending large universities in the US, South Korea and China in order to understand young consumers' attitudes, perceptions and behavioural intentions towards the consumption of environmentally sustainable textile and apparel products. In the aforementioned study, it was found that consumers' product knowledge positively affected their purchase intention. Here consumers' product knowledge is related to the organic cotton apparel rather than general environmental knowledge about designing and manufacturing processes of the apparel products.

4.4 Different Sources of Knowledge About Sustainability

The main sources of information that the consumers use for gathering knowledge about purchasing sustainable products include public education, peer influence and corporate marketing information (Mu et al. 2012). An analysis conducted for examining the influence of these three types of sources of knowledge about sustainability among Korean consumers revealed that public education is an effective source of knowledge for sustainable apparel consumption. It meant that the declarative knowledge, which describes the facts and processes included in the sustainable theory given in public education can shape the apparel consumer about sustainable purchasing.

Confirming the positive influence of education as a source of knowledge for sustainable apparel consumption, Sampson (2009) conducted a study among textile college students of a university in the USA; and found a strong significant relationship between consumer knowledge of environmentally friendly products and motivation to research, search, attitude and purchase green apparel products. Based on the findings, it was realized that consumer knowledge of environmentally friendly products increases, motivation to search and buy environmentally friendly products also increases. For evaluating the consumer knowledge about the green industry initiatives and green brands, the survey included the items such as how the students select environmentally friendly brands, which product categories offer environmental products, and consumer knowledge of retailers that carry green apparel items.

Improving the consumers' level of knowledge about sustainable apparel products, their production processes and the related standards can be achieved via information on product labels and packages, advertisements in different media such as brochures, magazines and television. That kind of information sharing and communication can enhance the consciousness of apparel consumer that may result in sustainable apparel purchasing (Tevel 2013).

A way of enhancing consumers' awareness about environmental concerns and triggering them towards sustainable products can be educating the consumers or make them to join and contribute to sustainable initiatives. Web resources of companies, retailers or brands giving information about sustainable products and their differences from other products may promote the purchase and use of sustainable apparel products (Kim 2010).

Labels on the sustainable apparel products is also an effective way of providing information about the products' sustainability and initiating them to purchase. Use of standardized eco-labelling systems may probably enhance the knowledge of consumers about sustainable apparel products in different concerns such as environmental and social (Tevel 2013).

The media, particularly social media recently, is inevitably one of the most important and effective way of influencing perception, awareness and knowledge of consumers about sustainable apparel. The mass media and social media has the ability to affect any change into a more sustainable society. This may be achieved via the messages and information provided with regard to eco-friendliness and sustainability (Sterling and Huckle 2014). Certainly, many people may believe that media is not responsible for educating the society in terms of sustainability; nonetheless, media is a very effective tool that can improve knowledge and awareness of society about sustainability to a great extent.

Tevel (2013) examined the role of knowledge in consumer willingness to purchase sustainable apparel and textile products for understanding the impact of environmental concern, attitude and knowledge on consumer willingness to purchase sustainable apparel and textiles, when knowledge being the main variable investigated. The results revealed that there is no significant correlation between knowledge and willingness to purchase sustainable apparel and textiles. However, further analysis showed that higher level of knowledge on sustainable apparel and textiles had higher willingness to purchase sustainable products.

Due to the growing importance of sustainability concern in overall business environment, the textile and apparel industry has also adapted the industrial practices towards a more sustainable approach. The transition of the industry should also be supported by the professionals equipped with the theoretical knowledge and skills of sustainability. Therefore, the sustainability related courses have been included into the curricula of undergraduate and graduate textile and apparel education programs. The courses equip the students to understand current and future need of sustainability and relating it to environmental, social and economic issues. An empirical research conducted among the participants in their third, fourth or fifth year of undergraduate education in the apparel and textile program at a higher education institution located in the Midwestern USA assessed their knowledge of social and environmental sustainability issues related to the industry and their apparel purchasing behaviour (Hiller Connell and Kozar 2012). The results of the study revealed that the integration of sustainability into undergraduate courses can successfully increase knowledge of the students about sustainability related social and environmental issues in the textile and apparel industry. In this way, students can gain ability to identify, evaluate, and analyse industry related sustainability issues. Besides, the study showed that despite the fact that knowledge may increase, there may not be a significant change in sustainable apparel purchasing behaviour. The two probable reasons behind could be either limited product availability or limited financial resources for purchasing sustainable apparel (Hiller Connell and Kozar 2012).

The information given by the corporates in fashion and apparel industry about environmental and social concerns have great influence on consumers' attitude and behavioural intention toward sustainable apparel products. The responses collected from the consumers in South Korea through an online survey indicated that apparel brands' marketing communication activities about environmental and social concerns impact consumers' attitude and behavioural intention toward sustainable fashion products (Kong et al. 2016). Indeed, it was reported by Park et al. (2014) that Korean consumers have trust and high expectations of companies' socially responsible activities.

However, this may not be so for the consumers all around the world. The empirical surveys conducted among Scandinavian consumers indicated that brand communications related to corporate social responsibility can be unconvincing when it is made by the brand itself. Besides, when such marketing communications of the companies regarding their corporate social responsibility are considered inappropriate by the consumer, they can be also believed to be doubtful (Morsing and Schultz 2006).

Kong et al. (2016) investigated the effect of four sustainable knowledge types including declarative knowledge, procedural knowledge, effectiveness knowledge and social knowledge on consumers' behavioural intention toward sustainable fashion products. "Declarative knowledge" refers to the understandings regarding facts and the semantic structure behind sustainability theory; whereas, procedural knowledge addresses possible actions, for example by increasing awareness about recycling plans (Kaiser and Fuhrer 2003). Effectiveness knowledge is related to the potential cost benefits from energy efficiency and effectiveness (Gardner and Stern 1996). Finally, social knowledge involves understanding others' motives and intentions (Ernst 1994; Ernst and Spada 1993) and it refers to actions in response to conventional norms and needs for social approval (Kaiser and Fuhrer 2003). The research results showed that effectiveness knowledge and social knowledge have positive impacts on Korean consumers' attitudes toward sustainable fashion products. Thus, action-based knowledge is necessary for referring consumers to the sustainable apparel product consumption. The results obtained by Kong et al. (2016) also showed that consumers do not rely on the knowledge they acquire from their peers.

Notwithstanding the extensiveness of green communities, the number of studies focusing on the social interactions or content creation by community members as a source of knowledge is quite limited. The overwhelming point of view in environmental behaviour of the consumer refers to the isolation of individuals' network and their implications in the consumer culture (Cervellon and Wernerfelt 2012).

However, today's one of the most efficient and commonly used knowledge sharing platform is online communities. Interaction of consumers through the online communities can be considered as a sustainable knowledge source channel. For investigating the knowledge sharing on green fashion and the sustainable supply chain within online communities and for analyzing the knowledge evolution with community members gaining in expertise Cervellon and Wernerfelt (2012) conducted a study by collecting discussions on green fashion from two green fashion forums over two periods (2007–2008 and 2010–2011), and a content analysis was performed. According to the findings of the study, as the community members gain expertise the objectivity of the knowledge they share increases and they may be sometimes based on scientific facts. From the retailer's side that offer sustainable apparel products to the market, the increase in the objectivity and accuracy of the knowledge shared between the members of any fashion related community has a power to encourages the development of sustainable processes in addition to the communication of the noticeable consequences for society (Cervellon and Carey 2011).

5 Emprical Study

The empirical study examined the effect of environmental sustainability knowledge, social sustainability knowledge and motivation for environmental responsibility on consumers in terms of their attitude towards green apparel brand and sustainable apparel purchasing intention.

5.1 Data Collection and Survey

The data for the empirical study was collected through an online survey conducted among participants in Turkey. The questionnaire of the survey was composed of three part with thirty-one questions having seven-point Likert scale, which ranged from 1 (strongly disagree) to 7 (strongly agree). In the first part, consumers' demographical and socio-economic information was gathered. This information involved age, gender, income level and educational attainment. The second part included the

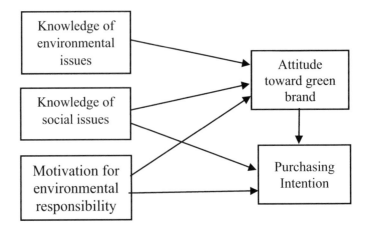

Fig. 2 The conceptual model

questions consumers' knowledge about environmental and social issues. In the third part, consumers' motivation for environmental responsibility, attitude toward green brand and purchasing intention was examined. The items examining the knowledge of environmental issues, knowledge of social issues, motivation for environmental responsibility, attitude toward green brand and purchasing intention were adapted from Kim (1995), Kozar and Hiller Connell (2010), Dunlap et al. (2000), Hartmann et al. (2005), Kim (1995) and Patney (2010) respectively.

The proposed hypotheses for the conceptual model that was demonstrated in Fig. 2 were given below.

H1   *Knowledge of environmental issues has a significant influence on consumer attitude toward green brand.*

H2   *Knowledge of social issues has a significant influence on consumer attitude toward green brand.*

H3   *Knowledge of social issues has a significant influence on consumer purchasing intention.*

H4   *Motivation for environmental responsibility has a significant influence on consumer attitude toward green brand.*

H5   *Motivation for environmental responsibility has a significant influence on consumer purchasing intention.*

H6   *Consumer attitude toward green brand has a significant influence on consumer purchasing intention.*

Table 1 Socio-demographic characteristics of the participants

	Item	N	%
Gender	Female	540	68
	Male	254	32
Age	18–23	183	23
	24–29	110	14
	30–35	56	7
	36-above	445	56
Level of education	High school graduate	110	14
	Attending undergraduate	199	25
	Bachelor graduate	342	43
	Enrolled in graduate school	32	4
	Graduate from graduate school	111	14
Income	Low	127	16
	Low to medium	214	27
	Medium to high	198	25
	High	255	32

5.2 Sample Profile

The data was collected from 794 participants. Among the participants, whose socio-demographic characteristics are given in Table 1, 68% was female, and 32% was male. 23% of the participants were between the ages of 18–23, 14% of the participants were between the ages of 24–29 and 7% of the participants were between the ages of 30–35. Finally, more than the half of the participants (56%) were in the age of 36 and above.

With regard to the education level, the profile of the participants can be summarized as follow: 44% of the participants were graduated with Bachelor Degree from university. 25% of them were still continuing undergraduate study. 14% of attendants had Master Degree. 4% were still continuing graduate study. 13% of participants were graduated from High School.

Four income categories, which were specified as low, low to medium, medium and medium to high income classes, were included in the questionnaire, and the results revealed that 16% of the participants were belonged to low income category, while 27% of the participants were belonged to low to medium income level category. 25% of the participants were observed to be belonged medium to high income category; and 32% of the participants were in high income category.

5.3 Statistical Analysis

The data were analysed statistically using IBM-SPPS and AMOS 21 (Analysis of Moment Structure) Program. The normality of the data was checked through the skewness and kurtosis of the data.

The scale reliability was assessed calculating the Cronbach's alpha. An exploratory factor analysis was performed using principal components with Varimax rotation in order to identify the constructs of the conceptual model. In the next step, the reliability and validity of the items composing the constructs were assessed using confirmatory factor analysis (CFA). Some modifications were made on the CFA checking the factor loadings and modification indices of error terms. The model fit was evaluated using the goodness of fit indices. The convergent validity of the conceptual model was tested by considering the average variance extracted (AVE) values and the composite reliability (CR) (Hair et al. 2006). The structural equation modelling (SEM) was employed to test the hypothesis and to evaluate the relationship between constructs. The statistical significance of the hypothesis was evaluated by comparing the actual value of critical ratio in AMOS to the preset critical value, the value of t at the desired two-tailed alpha or p-level (Clayton and Pett 2011), and it was taken as $t = 1.95$ at p level $= 0.05$.

5.4 Results

5.4.1 Scale Reliability and Exploratory Factor Analysis

The normality check of the data showed that none of the skewness values were greater than 3, and none of the kurtosis values were greater than 10. Therefore; no threat was found to normality (Kline 2005). The Cronbach's alpha measure of scale reliability was found to be (0.90), which is well above the generally acceptable value of 0.70 (Nunnally 1978).

The Kaiser-Meyer-Olkin (KMO) measure of sampling adequacy was found to be sufficiently large to justify factor analysis (0.916) (Keiser 1974). The result of Bartlett's test of sphericity was found significant. In the exploratory factor analysis, the items having factor loading more than 0.4 were retained. 23 items were grouped in five constructs, as shown in Table 2, and all together accounted for 71.974% of the total variance.

5.4.2 Confirmatory Factor Analysis Results

In order to evaluate the reliability and validity, Composite Reliability (CR), and Average Variance Extracted (AVE) were calculated for all of the constructs within the conceptual model. Displaying the degree to which a concept is represented by

Table 2 Exploratory factor analysis results

Item	Construct	Definition	Factors				
			1	2	3	4	5
20	PI	The likelihood that I would purchase from apparel retail brands engaged in CSR is very high	0.931				
21	PI	I would be willing to buy from apparel retail brands engaged in CSR activities	0.929				
23	PI	I have every intention to purchase from apparel retail brands engaged in CSR	0.926				
19	PI	I intend to recommend apparel retail brands engaged in CSR activities to my friends, family members, and co-workers	0.911				
22	PI	I am likely to purchase from apparel retail brands engaged in CSR in the future	0.907				
17	ATGB	Green product's environmental concern meets my expectations		0.860			
16	ATGB	I feel that green product's environmental claims are generally trustworthy		0.851			
15	ATGB	I feel that green product's environmental performance is generally dependable		0.800			
18	ATGB	Green products keep promises and responsibilities for environmental protection		0.758			
14	ATGB	I feel that green product's environmental reputation is generally reliable		0.701			
10	MFER	I should be responsible for protecting our environment			0.795		
11	MFER	Environmental protection starts with me			0.784		
9	MFER	Supporting environmental protection makes me feel as an environmentally responsible person			0.741		
12	MFER	I would say I am emotionally involved in environmental protection issues			0.600		
13	MFER	Supporting environmental protection makes me special			0.538		

(continued)

Table 2 (continued)

Item	Construct	Definition	Factors				
			1	2	3	4	5
6	KOS	Apparel manufacturers generally do not pay their employees at least the local minimum wage				0.820	
8	KOS	Apparel manufacturers generally provide hazardous workplaces for their employees				0.803	
7	KOS	Apparel manufacturers generally have their employees work more than 40 h per week				0.795	
5	KOS	Use of child labour is a practice among apparel manufacturers				0.756	
4	KOE	Phosphate-containing laundry detergents can be a source of water pollution					0.746
3	KOE	Textile dyeing and finishing processes use a lot of water					0.741
2	KOE	Air pollution can occur during some common dye processes of textiles					0.728
1	KOE	Chemical pollutants are produced during manufacturing of synthetic or manufactured fibres such as polyester					0.692

its indicators, composite reliability of the variables were found to be higher than 0.6 (Bagozzi and Yi 1988) getting the values 0.769 for "Knowledge of environmental issues", 0.828 for "Knowledge of social issues", 0.840 for "Motivation for environmental responsibility", 0.925 for "Attitude toward green brand" and finally 0.959 for "Purchasing intention".

The assessment of convergent validity was performed by calculating the Average Variance Extracted for all constructs and it was seen that they were approximate to or exceeded 0.5 which is an acceptable values (Bagozzi et al. 1991; Hair et al. 2006). They were calculated as 0.458 for "Knowledge of environmental issues", 0.549 for "Knowledge of social issues", 0.514 for "Motivation for environmental responsibility", 0.713 for "Attitude toward green brand" and finally 0.825 for "Purchasing intention".

Measuring the extent to which variables differ, discriminant validity was assessed by comparing the square roots of AVE values and with the correlations between the variable and all other variables (Teo et al. 2009). As the square roots of AVE values were found to be higher than the correlations between the variable and the other variables, it was confirmed that the discriminant validity was satisfied as seen in Table 3.

Table 3 The correlations between the variables and the square roots of AVE values

	Knowledge of environmental issues	Knowledge of social issues	Motivation for environmental responsibility	Attitude toward green brand	Purchasing intention
Knowledge of environmental issues	0.677				
Knowledge of social issues	0.547	0.741			
Motivation for environmental responsibility	0.496	0.298	0.717		
Attitude toward green brand	0.543	0.336	0.800	0.844	
Purchasing intention	0.122	0.040	0.228	0.287	0.909

Table 4 Evaluation of goodness of fit indices obtained from the analysis of conceptual model

Goodness of fit indices	Values used for comparison	Reference	Values obtained
Chi-square/degree of freedom	<3	Kline (2005)	2.124
Goodness of fit index (GFI)	≥0.9	Hair et al. (2006)	0.955
Adjusted goodness of fit index (AGFI)	≥0.8	Chau (1996)	0.941
Normalized fit index (NFI)	≥0.9	Hair et al. (2006)	0.966
Comparative fit index (CFI)	≥0.9	Hair et al. (2006)	0.981
Root mean square residual (RMR)	≤0.08	Hair et al. (2006), Brown and Cudeck (1992)	0.058
Root mean square of approximation (RMSEA)	≤0.08	Hair et al. (2006), Brown and Cudeck (1992)	0.038

The goodness of fit values in confirmatory factor analysis were observed to be within the recommended values as seen in Table 4.

Table 5 Results of hypothesis testing

Hypothesis		Estimates	p-value	Result
H1	Knowledge of environmental issues has a significant influence on consumer attitude toward green brand	0.216	0.000	Supported
H2	Knowledge of social issues has a significant influence on consumer attitude toward green brand	0.025	0.409	Not supported
H3	Knowledge of social issues has a significant influence on consumer purchasing intention	−0.062	0.253	Not supported
H4	Motivation for environmental responsibility has a significant influence on consumer attitude toward green brand	0.748	0.000	Supported
H5	Motivation for environmental responsibility has a significant influence on consumer purchasing intention	0.004	0.968	Not supported
H6	Consumer attitude toward green brand has a significant influence on consumer purchasing intention	0.403	0.000	Supported

5.4.3 Structural Equation Modelling Results

The results of the hypothesis testing are shown in Table 5. Based on the results, it was observed that H1, H4 and H6 were supported, whereas H2, H3 and H4 were not supported since the p-value was found to be below 0.05. The strength of the relationships for the supported hypothesis were estimated to be 0.748, 0.403 and 0.216 between the constructs "Motivation for environmental responsibility" and "attitude toward green brand", "attitude toward green brand" and "purchasing intention", and "Knowledge of environmental issues" and "attitude toward green brand". In addition, the multiple square correlations were found to be 0.669 for the construct "attitude toward green brand" and 0.086 for the construct "purchasing intention".

The strongest relationship was found between the motivation for environmental responsibility and consumer attitude toward green brand in terms of determining consumer approach toward sustainable apparel product. Caring about environmental issues and being more included into environmental protection activities influences the attitude of consumers toward green brands positively. This result is in agreement with the results obtained by Kumar and Ghodeshwar (2015), who concluded that consumers' decision to purchase a green product requires consciousness about environmental, individual and social consequences associated with green products.

The second important relationship was observed between the construct "attitude toward green brand" and "purchasing intention" with regard to determination of consumer's approach toward sustainable apparel products. Consumers' consideration about purchasing a sustainable apparel product shows their attitude toward green products and it determines their intention to purchase such products.

Last relationship, which is between the constructs "attitude toward green brand" and "knowledge of environmental issues" has significant but least influence regarding the determination of consumer approach toward sustainable apparel products. As the environmental knowledge and awareness increase, the attitude toward sustainable production affected positively and consistently.

As the fashion sense becomes more popular and apparel consumption increases excessively, production of textile material grew more profitable business. In this very competitive environment, social and environmental side effects became more distinct, and this made consumers more intended to embrace a new perception called sustainable fashion which adopts an ecologic and ethical way of production. This empirical study was made to lighten up the consumer's attitude toward sustainable products. According to studies made, several factors are chosen which affect this approach. These factors include knowledge of environmental issues, knowledge of social issues, drive for environmental responsibility, attitude toward green brand and intentions. These factors affect the approach to sustainable textile products directly as well as they affect each other in themselves.

5.5 Conclusions and Implications

This study proposed and validated a conceptual model based on primarily the effect of knowledge in terms of environmental issues and social issues with regard to sustainability on attitude of the consumers towards green brand and consumers' purchase intention of sustainable apparel products. Besides, the relationship between motivations for environmental responsibility and consumer attitude toward green brand was analysed.

The findings of the survey that was conducted among 794 consumers in Turkey revealed that knowledge of environmental issues had a significant and positive influence on consumer attitude toward green brand. In addition, motivation for environmental responsibility had a significant and positive influence on consumer attitude toward green brand, and consumer attitude toward green brand had a significant and positive influence on consumer purchasing intention.

The study also demonstrated that consumers' knowledge about social issues was not a predictor for consumer attitude toward green brand or consumer purchasing intention. This results suggests that the apparel manufacturers and retailers should emphasize their socially responsible practices through different sources of knowledge reaching to the consumer.

6 Recommendations for Future Study

Consumers can be conscious about environmental issues in some respects such as recycling daily use staffs and materials including paper, plastics and glass; or they can be sensitive to the energy conservation. Moreover, dietary habits of the consumers may be influenced towards the consumption of healthy foods; i.e. organic foods; easier with the awareness and increasing knowledge about sustainability when compared to the change of apparel consumption habits towards sustainable products. This is because sustainable apparel products are still perceived as out of fashion products, and they are far away from expressing the social status or lifestyle of the consumer.

Most of the studies regarding the investigation and evaluation of the relationship between consumer knowledge about sustainability concerns in the apparel industry and their attitude towards sustainable apparel consumption and purchasing were conducted among general consumer profiles without considering any market segmentation or consumer categories. Integrating the consumer knowledge based surveys with the consumers' clothing and purchasing purposes for different market segments would enable to understand which barriers are existing in front of purchasing sustainable apparel products from consumers' perspective. On the other hand, it will also be helpful for the business organizations to set appropriate strategies for becoming a more sustainable company and for promoting sustainable products.

References

Alpat, F. E. (2013). Yavaş Moda Nedir? *Akdeniz Sanat Dergisi, 4*(8).

Bagozzi, R. P., & Yi, Y. (1988). On the evaluation of structural equation models. *Journal of the Academy of Marketing Science, 16*(1), 74–94.

Bhardwaj, V., & Fairhurst, A. (2010). Fast fashion: Response to changes in the fashion industry. *The International Review of Retail, Distribution and Consumer Research, 20*(1), 165–173.

Bagozzi, R. P., Yi, Y., & Phillips, L. W. (1991). Assessing construct validity in organizational research. *Administrative Science Quarterly, 36*, 421–458.

Black, S. (2008). *Eco-chic: The fashion paradox*. London: Black Dog Publishing.

Browne, M. W., & Cudeck, R. (1992). Alternative ways of assessing model fit. *Sociological Methods Research, 21*(2), 230–258.

Can, O., & Ayvaz, K. M. (2017). Tekstil ve Modada Sürdürülebilirlik. *Tekstil, 1*(3), 110–119.

Capdevila, I. (2014). *How can living labs enhance the participants' motivation in different types of innovation activities?* Available https://ssrn.com/abstract=2502795 or http://dx.doi.org/10.2139/ssrn.2502795. Accessed Feb 15, 2018.

Carew, A. L., & Mitchell, C. A. (2002). Characterizing undergraduate engineering students' understanding of sustainability. *European Journal of Engineering Education, 27*(4), 349–361.

Carrigan, M., & Attalla, A. (2001). The myth of the ethical consumer–do ethics matter in purchase behaviour? *Journal of consumer marketing, 18*(7), 560–578.

Cervellon, M. C., & Carey, L. (2011). Consumers' perceptions of 'green': Why and how consumers use eco-fashion and green beauty products. *Critical Studies in Fashion & Beauty, 2*(1–2), 117–138.

Cervellon, M. C., & Wernerfelt, A. S. (2012). Knowledge sharing among green fashion communities online: Lessons for the sustainable supply chain. *Journal of fashion marketing and management: An international journal, 16*(2), 176–192.

Chau, P. Y. K. (1996). An empirical assessment of a modified technology acceptance model. *Journal of Management Information Systems, 13*(2), 185–204.

Chen, Y. S., & Chang, C. H. (2012). Enhance green purchase intentions: The roles of green perceived value, green perceived risk, and green trust. *Management Decision, 50*(3), 502–520.

Clark, H. (2008). SLOW + FASHION—An oxymoron—Or a promise for the future? *Fashion Theory, 12*(4), 427–446.

Clayton, M. F., & Pett, M. A. (2011). Modeling relationships in clinical research using path analysis Part II: Evaluating the model. *Journal for Specialists in Pediatric Nursing, 16*(1), 75–79.

Creyer, E. H., & Ross, W. T. (1997). Tradeoffs between price and quality: How a value index affects. *Journal of Consumer Affairs, 31*(2), 280–302.

D' Souza, C., Taghian, M., & Lamb, P. (2006). An empirical study on the influence of environmental labels on consumers. *Corporate communications: An international journal, 11*(2), 162–173.

Dickson, M. A. (2001). Utility of no sweat labels for apparel consumers: Profiling label users and predicting their purchases. *Journal of Consumer Affairs, 35*(1), 96–119.

Dickson, M. A. (1999). US consumers' knowledge of and concern with apparel sweatshops. *Journal of Fashion Marketing and Management, 3,* 44–55.

Dickson, M. A. (2000). Personal values, beliefs, knowledge, and attitudes relating to intentions to purchase apparel from socially responsible businesses. *Clothing and Textiles Research Journal, 18,* 19–30.

Dunlap, R. E., Van Liere, K. D., Mertig, A. G., & Jones, R. E. (2000). New trends in measuring environmental attitudes: measuring endorsement of the new ecological paradigm: A revised NEP scale. *Journal of social issues, 56*(3), 425–442.

Ernst, A. M. (1994). *Social knowledge as basis of a person's action in conflict situations.* Frankfurt: Lang.

Ernst, A. M., & Spada, H. (1993). Modeling agents in a resource dilemma: A computerized social learning environment. In D. Towne, T. de Jong, & H. Spada (Eds.), *Simulation-based experiential learning* (pp. 105–120). Berlin: Springer.

Eze, U. C., & Ndubisi, N. O. (2013). Green buyer behavior: Evidence from Asia consumers. *Journal of Asian and African Studies, 48*(4), 413–426.

Fletcher, K. (2007). Slow fashion. *The Ecologist, 37*(5), 61.

Gabrielli, V., Baghi, I., & Codeluppi, V. (2013). Consumption practices of fast fashion products: A consumer-based approach. *Journal of Fashion Marketing and Management, 17*(2), 206–224.

Gam, H. J., & Banning, J. (2011). Addressing sustainable apparel design challenges with problem-based learning. *Clothing and Textiles Research Journal, 29*(3), 202–215.

Garbie, I. (2016). *Sustainability in manufacturing enterprises: Concepts, analyses and assessments for industry 4.0.* Springer.

Gardner, G. T., & Stern, P. C. (1996). *Environmental problems and human behavior.* Allyn & Bacon.

Gürcüm, H. B., & Yüksel, C. (2012). Moda Sektörünü Yavaşlatan Eğilim: Eko Moda ve Moda'da Sürdürülebilirlik. 1. Uluslararası Moda ve Tekstil Tasarımı Sempozyumu.

Hair, J. F., Black, W. C., Babin, B. J., Anderson, R. E., & Tatham, R. L. (2006). *Multivariate data analysis* (6th ed.). Englewood Cliffs, N.J.: Prentice Hill.

Hartmann, P., Apaolaza Ibáñez, V., & Forcada Sainz, F. J. (2005). Green branding effects on attitude: Functional versus emotional positioning strategies. *Marketing Intelligence & Planning, 23*(1), 9–29.

Hiller Connell, K. Y., & Kozar, J. M. (2012). Sustainability knowledge and behaviors of apparel and textile undergraduates. *International Journal of Sustainability in Higher Education, 13*(4), 394–407.

https://apparelcoalition.org/higg-brand-tool/.

https://apparelcoalition.org/higg-facility-tools/.

https://apparelcoalition.org/higg-product-tools/.

https://apparelcoalition.org/the-sac/.

http://sustainability.hm.com/en/sustainability/about/about/ceo-message.html.

Hustvedt, G., & Bernard, J. C. (2008). Consumer willingness to pay for sustainable apparel: The influence of labelling for fibre origin and production methods. *International Journal of Consumer Studies, 32*(5), 491–498.

Hustvedt, G. (2006). *Consumer preferences for blended organic cotton apparel.* (Dissertation, Kansas State University).

Hustvedt, G., & Dickson, M. A. (2009). Consumer likelihood of purchasing organic cotton apparel: Influence of attitudes and self-identity. *Journal of Fashion Marketing and Management: An International Journal, 13*(1), 49–65.

James, P. (2014). *Urban sustainability in theory and practice: Circles of sustainability.* Routledge.

Jung, S., & Jin, B. (2014). A theoretical investigation of slow fashion: Sustainable future of the apparel industry. *International journal of consumer studies, 38*(5), 510–519.

Kagawa, F. (2007). Dissonance in students' perceptions of sustainable development and sustainability: Implications for curriculum change. *International Journal of Sustainability in Higher Education, 8*(3), 317–338.

Kang, J., Liu, C., & Kim, S.-H. (2013). Environmentally sustainable textile and apparel consumption: the role of consumer knowledge, perceived consumer effectiveness and perceived personal relevance. *International Journal of Consumer Studies, 37*(4), 442–452. https://doi.org/10.1111/i jcs.12013.

Keiser, H. F. (1974). An Index of factorial simplicity. *Psychometrika, 39,* 31–36.

Kaiser, F. G., & Fuhrer, U. (2003). Ecological behavior's dependency on different forms of knowledge. *Applied Psychology, 52*(4), 598–613.

Keller, K. L. (1993). Conceptualizing, measuring, and managing customer-based brand equity. *Journal of Marketing, 57*(1).

Kim, H. S. (1995). *Consumer response toward apparel products in advertisements containing environmental claims.*

Kim, Y. K. (2010). Standards and certifications for sustainable textile & fashion industries. *Fashion Information and Technology, 7.*

Kim, H. S., & Damhorst, M. L. (1998). Environmental concern and apparel consumption. *Clothing & Textiles Research Journal, 16,* 126–133.

Kim, S., Littrell, M. A., & Paff Ogle, J. L. (1999). Academic papers: The relative importance of social responsibility as a predictor of purchase intentions for clothing. *Journal of Fashion Marketing and Management: An International Journal, 3*(3), 207–218.

Kirsi Niinimäk. https://shop.aalto.fi/media/attachments/1ee80/SustainableFashion.pdf.

Kline, R. B. (2005). *Principles and practice of structural equation modelling.* Guilford Press.

Kong, H. M., Ko, E., Chae, H., & Mattila, P. (2016). Understanding fashion consumers' attitude and behavioral intention toward sustainable fashion products: Focus on sustainable knowledge sources and knowledge types. *Journal of Global Fashion Marketing, 7*(2), 103–119.

Kozar, J. M., & Hiller Connell, K. Y. (2010). Socially responsible knowledge and behaviors: Comparing upper- versus lower-classmen. *College Student Journal, 44,* 279–293.

Kumar, P., & Ghodeswar, B. M. (2015). Factors affecting consumers' green product purchase decisions. *Marketing Intelligence & Planning, 33*(3), 330–347.

Lin, S. H. (2009). Exploratory evaluation of potential and current consumers of organic cotton in Hawaii. *Asia Pacific Journal of Marketing and Logistics, 21*(4), 489–506.

MacArthur, E. (2014). Towards the circular economy: Accelerating the scale-up across global supply chains. In: *World Economic Forum*, Geneva. https://doi.org/10.1162/108819806775545321.

Mangır, A. F. (2016). Sürdürülebilir Kalkınma için Yavaş ve Hızlı Moda. *Selçuk Üniversitesi Sosyal Bilimler Meslek Yüksek Okulu Dergisi, 19,* 143–154.

Morsing, M., & Schultz, M. (2006). Corporate social responsibility communication: Stakeholder information, response and involvement strategies. *Business Ethics: A European Review, 15*(4), 323–338.

Mu, J. W., Lee, S. S., & Ryu, M. H. (2012). Consumption awareness according to information search and consumer education for green consumption: Comparative study between Korea and China. *Journal of Korean Home Management Association, 30,* 29–44.

Niinimäki, K., & Hassi, L. (2011). Emerging design strategies in sustainable production and consumption of textiles and clothing. *Journal of Cleaner Production, 19*(16), 1876–1883.

Niinimäki, K. (2013). *Sustainable fashion: new approaches.* Aalto University.

Nike. (2010). *Nike releases environmental design tool.* Available https://news.nike.com/news/nike-releases-environmental-design-tool-to-industry. Accessed Mar 02, 2018.

Nunnally, J. C. (1978). *Psychometric theory.* New York: McGraw-Hill.

Outdoor Industry Association. (2014). *OIA Social Responsibility Toolkit 2.0.* Available https://outdoorindustry.org/wp-content/uploads/2015/04/OIA-Social-Responsibility-Toolkit.pdf. Accessed Mar 02, 2018.

Page, G., & Fearn, H. (2005). Corporate reputation: What do consumers really care about? *Journal of Advertising Research, 45*(3), 305–313.

Pagiaslis, A., & Krontalis, A. K. (2014). Green consumption behavior antecedents: Environmental concern, knowledge, and beliefs. *Psychology & Marketing, 31*(5), 335–348.

Park, S., Kim, Y., Lee, U., & Ackerman, M. (2014, Sept). Understanding localness of knowledge sharing: a study of Naver KiN'here'. In: *Proceedings of the 16th international conference on Human-computer interaction with mobile devices & services* (pp. 13–22). ACM.

Patney, M. B. (2010). *Indian consumers and their mall patronage: Application of cultural-self and the theory of planned behavior to patronage intentions.* Iowa State University.

Peattie, K. (2010). Green consumption: behavior and norms. *Annual review of environment and resources, 35.*

Pulse of the Fashion Industry. (2017). Available http://globalfashionagenda.com/wp-content/uploads/2017/05/Pulse-of-the-Fashion-Industry_2017.pdf. Accessed Mar 15, 2018.

Sampson, L. K. (2009). *Consumer analysis of purchasing behavior for green apparel (MSc Thesis).* North Carolina State University.

Saricam, C. & Erdumlu, N. (2010). The impact of knowledge on the purchasing decision of environmentally friendly apparel. In: *41st International Symposium on Novelties in Textiles,* 27–29 May (2010), Ljubljana, Slovenia.

Sterling, S., & Huckle, J. (Eds.). (2014). *Education for sustainability.* Routledge.

Stephens, S. H. (1985). *Attitudes toward socially responsible consumption: Development and validation of a scale and investigation of relationships to clothing.* (Doctoral dissertation). Virginia Polytechnic Institute and State University, Blacksburg, VA.

Stevenson. (2013). *In textiles, green is the new black.* Available https://www.itma.com/docs/default-source/news/itma_sustainability_bulletin_issue_1_2013_en.pdf?sfvrsn=2. Accessed Feb 26, 2018.

Teo, T., Lee, C. B., Chai, C. S., & Wong, S. L. (2009). Assessing the intention to use technology among preservice teachers in Singapore and Malaysia: A multigroup invariance analysis of the technology acceptance model. *Computers & Education, 53,* 1000–1009.

Tevel, A. (2013). *The affect of knowledge on consumer willingness to purchase sustainable apparel and textiles.* (Doctoral dissertation, California State University, Northridge).

The Ecologist January. (1972). Available https://theecologist.org/2012/jan/27/ecologist-january-1972-blueprint-survival. Accessed Feb 04, 2018.

The Ellen MacArthur Foundation. (2014). Towards the Circular Economy, Vol. 3: Accelerating the Scale-up Across Global Supply Chains.

Thiele, L. P. (2014). *Sustainability.* Cambridge: Polity.

Thompson, C. J. (1995). A contextualist proposal for the conceptualization and study of marketing ethics. *Journal of Public Policy & Marketing,* 177–191.

Tischner, U., & Charter, M. (2001). *Sustainable solutions: Developing products and services for the future.* Greenleaf.

Türkmen, N. (2009). Sustainability and Transformation from the Aspect of Textiles and Fashion Design, Mimar Sinan Fine Arts University, Institute of Social Sciences, Proficiency in Arts, İstanbul.

Yadav, R., & Pathak, G. S. (2016). Young consumers' intention towards buying green products in a developing nation: Extending the theory of planned behavior. *Journal of Cleaner Production, 135,* 732–739.

Wiersum, K. F. (1995). 200 years of sustainability in forestry: Lessons from history. *Environmental Management, 19,* 321–329.

Willard, B. (2012). *The new sustainability advantage: Seven business case benefits of a triple bottom line.* New Society Publishers.

World Commission on Environment and Development-Brundtland Report (1987).

Printed in the United States
By Bookmasters